T4-APU-452

Applications Manual for Paint and Protective Coatings

Applications Manual for Paint and Protective Coatings

A GUIDE TO TYPES OF COATINGS
METHODS OF SURFACE PREPARATION
AND HAND APPLICATION TECHNIQUES

William F. Gross

Binks Research and Development Corporation
Boulder, Colorado

McGRAW-HILL BOOK COMPANY

New York St. Louis San Francisco Düsseldorf
London Mexico Panama
Sydney Toronto

APPLICATIONS MANUAL FOR PAINT AND PROTECTIVE COATINGS

 Library of Congress Catalog Card Number 71-107290

07-024970-9

1234567890 MAMM 7543210

Foreword

Previously written paint manuals tend to limit their coverage to a narrow field of interest, which is then treated in far too great detail for the average paint applicator. A need has been expressed for a reference and working manual which covers all subjects concerned with the hand application of paint, but presents only the essential details under each subject.

William F. Gross has the broad background necessary for the preparation of such a manual. As a chemical engineer, he has acquired an extensive knowledge of coating formulation and application techniques. While staff corrosion engineer for a major overseas oil company, he subsequently gained considerable experience in evaluating paints and coatings under various exposure conditions.

At present, Dr. Gross is associated with Binks Research and Development Corporation, which is a wholly owned subsidiary of Binks Manufacturing Company, Chicago—a prominent fabricator of paint application equipment.

Successful paint application requires a thorough knowledge of paint characteristics, surface preparation and application techniques, and the customer's service requirements. It is fortunate that Dr. Gross has had extensive experience in these fields, and so can combine the specialized knowledge of each into one manual.

The "Applications Manual for Paint and Protective Coatings" has been prepared with three functions in mind. It is intended to serve as

1. A reference manual regarding types and characteristics of paints; color coding for safety and identification of contents; and health and safety considerations
2. An instruction manual regarding surface preparation; paint preparation; and various techniques in the hand application of paint
3. A working manual presenting recommended coating systems for various service conditions, together with specific application instructions

In recent years we have witnessed rapid advances in both paint formulation and paint application equipment. Increasing complexity and continuous refinements in paints and protective coatings have necessitated new developments in application equipment and techniques. To those who are actively involved in this area, this book will be of great value as it answers an obvious need for simplified and dependable information in a complex field.

Burke B. Roche, president
Binks Manufacturing Company
Chicago, Illinois

Contents

Applications Manual for Paint and Protective Coatings

Chapter 1

Types of Coatings and Their Characteristics

Introduction

Organic coatings are composed of pigments suspended in a vehicle. The vehicle or carrier consists primarily of a resinous binder dissolved in solvents, together with small quantities of driers, plasticizers, and stabilizers as required. As the paint film dries, these vehicles change from a liquid to the solid film by one or more of several mechanisms:

1. Evaporation of solvents
2. Oxidation (of a drying oil)
3. Polymerization through application of heat, addition of a catalyst, or combination of reactive components.

The pigments contribute such properties as inhibition of the metal surface (red lead and zinc chromate), reinforcement of the film, stabilization against deterioration by sunlight, controlled chalking (titanium dioxide), and color. Clear coatings (varnishes, lacquers, and shellac) are not pigmented.

Inorganic coatings such as the zinc silicates also may be thought of as a pigment suspended in a vehicle. In this case, the pigment consists of metallic zinc dust, and the vehicle is a blend of water-soluble silicates.

Terminology sometimes is confusing. "Paint" usually refers to materials applied primarily for their decorative effect, and any protection received is of secondary importance. The term "coating" or "protective coating," on the other hand, generally is reserved for use when protection is the primary consideration.

Paints and coatings often are classified according to the type of vehicle employed. The following discussion presents the principal characteristics of the more important generic types of paints and coatings. This is followed on page 31 by a brief review of pigments: their characteristics and their effect on paint life.

Drying Oil (Oleoresinous) Paints

All paints grouped in this classification have in common the use of an unsaturated or "drying" oil in the binder. The drying oil may have synthetic resins added in solution in the vehicle as with oil-base paints, or the oil may be chemically reacted with the resins during "cooking" of the varnish, as with alkyds, phenolic varnishes, and epoxy esters.

All paints of this class dry initially by solvent evaporation, followed by curing of the paint film through addition of oxygen from the air at points of unsaturation in the oil. Metallic driers often are added to speed the drying time by accelerating the addition of oxygen.

Because of the presence of the unsaturated oil in the binder, drying oil paints generally are unsuitable for immersion service. They also show little resistance to attack by acids and alkalies (caustics).

Oil-base paint: Oil-base paints are based on a refined, boiled, heat-bodied, or chemically treated unsaturated oil (usually linseed, tung, fish, soybean, or dehydrated castor oil). Oil-base films dry initially by solvent evaporation, followed by the slower addition of oxygen of the air at points of unsaturation in the oil.

The binder in these paints consists either of drying oils alone or blends of these with natural or synthetic resins. Metallic driers are added to speed the drying time by accelerating the addition of oxygen at the points of unsaturation in the oil. Nevertheless, this is a slow process, and therefore oil-base paints should be permitted two days' dry time between coats.

The film formed is flexible and has higher permeability to water vapors and other gases than do films of the synthetic resin paints. Drying oil paints exhibit very good weathering properties. Their recommended service is limited to atmospheric exposure under mildly humid and corrosive conditions in the absence of chemical, alkali, and acid fumes. They show a short life in immersion service.

An outstanding property of the oil-base paints is their ability to wet and adhere to steel, which accounts for their widespread use as primers. This property is particularly important in instances where it is not possible to remove all rust before painting. Furthermore, their permeability makes them good binders in house paints for use on wood, where without some permeability the paint would blister and peel.

A suitable cleaning solvent and thinner for these paints is mineral spirits or painter's naphtha. The use of turpentine may be advantageous in some cases. Boiled linseed oil often is added to reduce the concentration of pigments, as is desired in the first coat (primer sealer) applied to bare wood.

No special surface preparation is required. Any practical application method using brush, spray, or roller may be employed. Representative coverage is 350 to 450 sq ft per gal per coat, providing 2- to 2½-mils dry film thickness.

Representative spraying conditions are:

1. *Air Atomization*
   FLUID PRESSURE: 25 to 30 psig
   ATOMIZING AIR PRESSURE: 30 to 40 psig
   FLUID ORIFICE SIZE: 0.070 to 0.086 in.
   AIR NOZZLE AIR RATE: 12 to 14 cfm at 50 psig
2. *Airless Spray*
   FLUID PRESSURE: 1,600 to 2,000 psig
   FLUID ORIFICE SIZE: 0.015 to 0.018 in.

Alkyd paint and alkyd varnish: Most alkyd resins for coatings are made by formulating certain polybasic acids (phthalic, fumaric, maleic, succinic, etc.) with polyhydric alcohols (glycerol, pentaerythritol, glycols, etc.), reacted or cooked together with varying quantities of drying oils which impart flexibility to the resin. Like oil-base paints, alkyds dry by reacting with atmospheric oxygen. The alkyd vehicle free of pigments, and with some adjustment in the relative quantity of the components, forms alkyd varnish.

The greater the oil content in the resin, the more flexible the coating will be; however, it also follows that the greater the oil content or "length," the more the film will resemble an oil-base paint. In general, the various oil-modified alkyds are referred to as short, medium, or long oil alkyds.

Alkyd paints are faster drying than are oil-base paints, the drying time depending upon the amount of oil incorporated during formulation. Alkyds with shorter oil lengths will have quicker drying times; however, they also exhibit less steel wetting ability and so require better surface preparation. Short oil alkyds are commonly employed as drum enamels because of their quick drying time.

The outstanding properties of alkyds are their good resistance to weathering, good gloss retention, durability, stability to sunlight, and somewhat greater resistance to water (moisture) and alkalies than is shown by oil base paint films.

Alkyd paints are more widely used on metal and wood products than any other paint system. Their major applications are in industrial plants for protection of steel surfaces from atmospheric corrosion and as an all-purpose paint for wood and metals in atmospheric exposure (medium oil alkyds). They are not suitable for use in immersion service.

Suitable cleaning solvents and thinners are the same as for oil-base paints: mineral spirits, painter's naphtha, or turpentine. The addition of 10 to 20 percent aromatic solvent (toluene) to the mineral spirits will improve its action in thinning and cleaning.

Any practical application method using brush, spray, or roller may be employed. Representative coverage is 350 to 400 sq ft per gal per coat, providing 1- to 1½-mils dry film thickness.

Representative spraying conditions are:

1. *Air Atomization*

   FLUID PRESSURE: 25 to 30 psig
   ATOMIZING AIR PRESSURE: 30 to 40 psig
   FLUID ORIFICE SIZE: 0.070 to 0.086 in.
   AIR NOZZLE AIR RATE: 12 to 14 cfm at 50 psig

2. *Airless Spray*

   FLUID PRESSURE: 1,600 to 2,000 psig
   FLUID ORIFICE SIZE: 0.015 to 0.018 in.

Phenolic varnish paint and phenolic varnish: One of the first synthetic resins to find wide use in coatings was the oil-modified phenolic resin, formed by reacting a substituted phenol with formaldehyde and a drying oil. These air-drying phenolic varnish films dry by solvent evaporation and oxygen adsorption in the same manner as do oil-base and alkyd paint films. (They are not to be confused with the heat-cured and catalytically cured phenolic coatings which are discussed later.) As with alkyds, in the absence of pigments and with minor adjustment in composition, the vehicle forms phenolic varnish.

Phenolic varnish paints have better chemical and water resistance than do oil-base and alkyd paints. However, they are less resistant to the effects of ultraviolet light, tending to darken slightly with age and to chalk more rapidly. Phenolic varnish coatings are employed where durable protection against atmospheric corrosion is required, especially in the presence of moisture or corrosive fumes. Consequently, they are commonly used for marine varnishes and paints. Phenolic paints are fast drying, with the dry time again depending upon the amount of drying oil incorporated.

Suitable cleaning solvents and thinners are mineral spirits or enamel thinner, with added aromatic solvents (toluene and xylene) to improve solvency.

A good surface preparation is recommended before paint application. Any practical application method using brush, spray, or roller may be employed. Representative coverage is 250 to 350 sq ft per gal per coat, providing 1½- to 2-mils dry film thickness.

Representative spraying conditions are:

1. *Air Atomization*

   FLUID PRESSURE: 25 to 30 psig
   ATOMIZING AIR PRESSURE: 30 to 40 psig
   FLUID ORIFICE SIZE: 0.070 to 0.086 in.
   AIR NOZZLE AIR RATE: 12 to 14 cfm at 50 psig

2. *Airless Spray*
   FLUID PRESSURE: 1,800 to 2,000 psig
   FLUID ORIFICE SIZE: 0.015 to 0.018 in.

Epoxy ester paint: Epoxy resins are formed by reaction of epichlorohydrin with bisphenol A. The resulting resin is then further reacted with a drying oil (or fatty acid) such as linseed oil, to form epoxy esters. Epoxy esters closely parallel alkyd resins and phenolic varnishes in vehicle properties and drying requirements.

Paints based on these resins are somewhat more resistant to chemical fumes and salty atmospheres but have less gloss retention than do alkyds and phenolic varnish paints. They are generally used for protection against corrosive atmospheres since they exhibit very good resistance to a wide range of corrosive agents, oil, and moisture.

Epoxy esters find wide application as industrial maintenance paints because of their broad resistance to corrosive atmospheres, their easy application, and the low order of solvency of solvents employed. This means they may be applied over existing paints without causing lifting or blistering.

In common with other paints containing drying oils, epoxy ester paints are not suitable for water-immersion service.

Suitable cleaning solvents and thinners are the same as for phenolic varnishes: mineral spirits or enamel thinner, with added aromatic solvents (toluene and xylene) to improve solvency.

Good surface preparation is required before paint application. Any practical method using brush, spray, or roller may be employed. Representative coverage is 350 to 400 sq ft per gal per coat, providing 1- to 1½-mils dry film thickness.

Representative spraying conditions are:

1. *Air Atomization*
   FLUID PRESSURE: 25 to 30 psig
   ATOMIZING AIR PRESSURE: 30 to 40 psig
   FLUID ORIFICE SIZE: 0.070 to 0.086 in.
   AIR NOZZLE AIR RATE: 12 to 14 cfm at 50 psig

2. *Airless Spray*
   FLUID PRESSURE: 1,600 to 2,000 psig
   FLUID ORIFICE SIZE: 0.015 to 0.018 in.

Precured, Solvent Dry Paints

All paints grouped in this classification have in common the use of fully cured resins as the binder. Since the paint vehicles consist of these resins dissolved in a suitable solvent, the applied paint films will of course be attacked and dissolved when contacted by their respective solvents.

Because the resins are fully cured or polymerized before application, paints and coatings of this classification usually exhibit excellent resistance to water and chemicals. Consequently, they are commonly employed in immersion service.

Chlorinated rubber paint: The chlorinated rubber resin is obtained by treating natural rubber with chlorine. The resulting material is a hard, brittle resin and must be plasticized to make elastic paint films. In addition, stabilizers are added to chlorinated rubber paints for exterior service to prevent deterioration by ultraviolet light. The chlorinated rubber resins are completely saturated, and these paint films dry solely by solvent evaporation.

Chlorinated rubber paints are formulated to have a range of properties depending upon the service condition intended—such as application to steel, concrete, or wood, and for atmospheric or immersion service. However, for the paint to exhibit the desired characteristics of chlorinated rubber, the chlorinated rubber content of the resin binder should be at least 50 percent. Because of the high chlorine content of the resin, quality chlorinated rubber paints do not support combustion.

Chlorinated rubber paints have very good adhesion to wood and concrete surfaces. When applied to steel, primers containing rust inhibitive pigments generally are employed, with drying oils or alkyd resins added for better adhesion.

Chlorinated rubber paints have excellent chemical resistance and are used in direct contact with acids or alkalies, or for severe fume conditions. Their outstanding properties are low permeability to water vapor (good resistance to water penetration) and outstanding resistance to alkalies. This combination of properties leads to their widespread use on concrete, especially for immersion service such as in swimming pools. They also resist wear when used as floor paints on concrete or wood. Since the resin is odorless and tasteless in addition to being unaffected by water, chlorinated rubber paints also are commonly employed for protection of drinking-water tanks.

Chlorinated rubber paints should not be applied in combination with other types of paints or coatings, because of incompatibility problems—principally due to the strong solvents employed. Furthermore, they must be applied to dry surfaces because of their low permeability to moisture. Any dampness on the surface will cause film blistering. Suitable cleaning solvents and thinners are aromatic hydrocarbons (toluene and xylene) and aromatic-type enamel thinners.

Good surface preparation is required before paint application, and the surface must be dry. These paints usually are applied by brushing. Spray application is difficult due to cobwebbing, and therefore these paints require greater than average skill in spraying.

Representative coverage is 300 to 350 sq ft per gal per coat, providing 1- to 1½-mils dry film thickness.

In general, spray application of chlorinated rubber paints is not recommended. The following spray conditions apply to coatings especially formulated for spraying:

1. *Air Atomization*
   FLUID PRESSURE: 8 to 12 psig
   ATOMIZING AIR PRESSURE: 50 to 65 psig
   FLUID ORIFICE SIZE: 0.070 to 0.078 in.
   AIR NOZZLE AIR RATE: 12 to 14 cfm at 50 psig

2. *Airless Spray*
   FLUID PRESSURE: 2,000 to 2,500 psig
   FLUID ORIFICE SIZE: 0.011 to 0.015 in.

Vinyl and vinyl-acrylic paint: The main constituents forming the binder of these coatings are polyvinyl chloride copolymerized with a minor portion of polyvinyl acetate. The resulting resin is tough and flexible, and so plasticizers are not required. Since it is completely saturated and polymerized, the paint film dries solely by solvent evaporation.

Vinyl coatings have outstanding durability and excellent resistance to acids, alkalies, chemicals, and salt water. They are superior to chlorinated rubber in resistance to oils and fats. Like chlorinated rubber, vinyl paints will not support combustion.

Their adherence is poor unless special primers are used. The vinyl-butyral wash primer was developed especially to improve vinyl adherence. Wash primers contain phosphoric acid which reacts with the sandblasted steel surface to which it is applied, producing an iron phosphate surface overcoated with a tightly bonded thin coat of vinyl-butyral resin. Other primers formulated for increased adhesion combine vinyls with resins such as alkyds.

The usual vinyl coating has the lowest solid content of any of the commonly used coatings. Consequently, thin coats are applied and multiple coats are necessary to build up adequate film thickness. This disadvantage is partially offset, however, by quick drying characteristics, permitting the application of several coats in a day. Application by hot spray increases the film thickness per coat, decreases drying time to a few minutes, and essentially eliminates pinholing.

Vinyls are probably the most widely used coating for stationary and mobile marine equipment and for protection of water tank interiors. They exhibit poor heat resistance. Since they dry only by solvent evaporation, they can be redissolved in the same solvents: ketones, aromatic hydrocarbons, and esters. They are fast drying and may be recoated within 2 hr.

Acrylic resins are sometimes added to improve brushing and gloss retention, with some sacrifice of the normal chemical resistance. Vinyl acrylics are widely used as an exterior paint in humid atmospheres.

High build (high solids content) vinyl mastic coatings are available. These may be applied in thick films (up to 5 to 6 mils) per coat. Because they dry only by solvent evaporation, they tend to be porous. Consequently, one or more seal coats are commonly applied over these mastics.

The polyvinyl acetate resin alone (without the polyvinylchloride resin) is commonly employed as a water-emulsion paint. These polyvinyl acetate water-base paints are discussed later in this section.

A very good sandblasted surface is required for the application of vinyl paints. Any practical application method using brush, spray, or roller may be employed; however, brushing is difficult due to the fast drying property of vinyl paint. Spraying is the preferred method of application. Solvent and equipment cleaners are ketones: methyl ethyl ketone (MEK) or preferably methyl isobutyl ketone (MIBK). Representative coverage is 200 to 250 sq ft per gal per coat, providing ¾- to 1-mil dry film thickness.

Representative spraying conditions are:

1. *Air Atomization*
   FLUID PRESSURE: 15 to 25 psig
   ATOMIZING AIR PRESSURE: 60 to 90 psig
   FLUID ORIFICE SIZE: 0.059 to 0.070 in.
   AIR NOZZLE AIR RATE: 14 to 16 cfm at 50 psig

2. *Airless Spray*
   FLUID PRESSURE: 1,900 to 2,100 psig
   FLUID ORIFICE SIZE: 0.013 to 0.015 in.

Silicone and silicone-alkyd paint: Silicone resins, because of their thermal stability, are used in formulating heat-resistant point. These resins are quite expensive, resulting in a paint cost averaging from $15 to $20 a gallon.

Aluminum-pigmented silicone paints perform well up to about 1,000°F. They are widely used on mufflers, furnace stacks, and similar hot surfaces. The silicone resin burns off at around 700°F, leaving the aluminum flakes physically adhering to the surface. However, as long as the stack surface stays warm, this adhesion is sufficiently strong to withstand normal weathering for 18 to 24 months.

No more than two coats of silicone coating should be applied, with a heat cure recommended after the first coat. Thin coats of ¾- to 1-mil thickness per coat provide the best service.

Although zinc-pigmented silicone primers are sometimes used for the first coat, experience has shown that superior results are obtained by applying a zinc silicate primer as the first coat on the sandblasted surface.

The silicone-alkyd paints were developed to fill the need for a glossy finish on moderately hot surfaces (up to 500°F). Silicone-alkyd paints require a primer over which one or two topcoats are applied in the color desired.

Suitable cleaning solvents and thinners are toluene, xylene, and enamel thinner. Surface preparation must be by sandblasting. Silicone paints may be applied by brush, but recommended application is by spraying. Representative coverage is 350 to 400 sq ft per gal per coat, providing ¾- to 1-mil dry film thickness.

Representative spraying conditions are:

1. *Air Atomization*
   FLUID PRESSURE: 10 to 20 psig
   ATOMIZING AIR PRESSURE: 40 to 70 psig
   FLUID ORIFICE SIZE: 0.040 to 0.052 in.
   AIR NOZZLE AIR RATE: 13 to 15 cfm at 50 psig
2. *Airless Spray*
   FLUID PRESSURE: 1,500 to 1,800 psig
   FLUID ORIFICE SIZE: 0.011 to 0.015 in.

Acrylic and nitrocellulose lacquers and enamels: These high gloss finishes are discussed together because of similar properties and usages. The resins of both lacquers are fully cured, and so the films dry by solvent evaporation only.

The acrylic lacquers are solutions of substituted polyacrylic and acrylate resins together with plasticizing resins. They dry quickly to provide a durable high gloss finish. Their outstanding properties are resistance to chalking and permanence of color and gloss. Their chemical, acid, and alkali resistance is poor, although recent formulations show improved chemical resistance.

They have become accepted as the present-day automobile finish, supplanting short-oil alkyds previously used. Because of their durable gloss during atmospheric exposure, acrylic lacquers also are finding wide usage as a finish coat for epoxies and vinyls, to provide improved appearance for longer periods of time.

The nitrocellulose lacquers are made from nitrocellulose materials blended with plasticizers and resins such as alkyds and polyacrylics. They dry quickly in air, are tough, and have good film strength. Clear coatings provide excellent appearance indoors; pigmented coatings for outdoor use have fair durability and good gloss and color retention. They are widely used, principally as decorative coatings for metal products for indoor use.

A suitable cleaning solvent and thinner for both acrylic and nitrocellulose lacquers is lacquer thinner, which is a blend of esters, ketones, and cyclic alcohols. The highly flammable nature of the solvents requires special attention in addition to normal safety precautions.

Recommended application for lacquers is by air spraying. Because a fine finish is desired when applying lacquers, application by air spray is recommended over brushing or airless spray. Brushing may leave brush marks, and airless spray produces a coarser finish. Average coverage per gallon is 450 to 500 sq ft, providing a coat of ¾- to 1-mil thickness.

Representative spraying conditions are:

1. *Air Atomization*
   FLUID PRESSURE: 5 to 10 psig
   ATOMIZING AIR PRESSURE: 50 to 60 psig
   FLUID ORIFICE SIZE: 0.040 to 0.052 in.
   AIR NOZZLE AIR RATE: 14 to 15 cfm at 50 psig
2. *Airless Spray*
   FLUID PRESSURE: 1,300 to 1,600 psig
   FLUID ORIFICE SIZE: 0.011 to 0.013 in.

Bituminous coatings (cold-applied): The solvent cutback asphalt and coal-tar coatings are discussed together because of similar properties and usages. The asphalt and coal-tar pitches are fully "cured," and so these films dry by solvent evaporation only.

The bituminous pitches are dissolved in suitable solvents: aromatic and coal-tar solvents for coal tar; aliphatic solvents such as mineral spirits for asphalt. To increase film thickness per coat, inert fillers often are incorporated. A popular type of cold-applied coating is a highly thixotropic, filled coal-tar mastic. The thick pastelike consistency is reduced by vigorous stirring until proper application viscosity is reached. Additional stirring may thin the material more than desired. This thixotropic type of coating should be thinned only mechanically—never with solvents.

Water emulsion forms of these coatings also are available; these exhibit special properties which are mentioned later. Clays are commonly employed as the dispersing agent.

The bituminous coatings find wide application as a low cost, highly effective heavy-duty industrial coating in corrosive exposures, wherever their black color is not objectionable. They are tough and durable, and exhibit excellent adhesion to both steel and concrete.

The coal-tar solvent cutback coatings are highly impermeable to moisture and are outstanding in their resistance to water deterioration. They also show very good resistance to petroleum oils, weak acids, alkalies, and salts, but not to hydrocarbon solvents and vegetable oils. Their principal use is for protection of buried or submerged steel and concrete. However, only selected coal-tar coatings, which have been processed to eliminate phenol tastes and odors, can be used in drinking-water tanks.

The coal-tar compositions, on prolonged exposure to weathering and direct sunlight, gradually lose a portion of their natural plasticizer by slow evaporation. Such volatilization causes shrinkage, which becomes evident through surface checking or "alligatoring." Shrinkage is not noticeable with the asphalt-base materials and is completely absent in water-dispersion coatings, probably because of the clay binder. The latter are often employed as a topcoat over bituminous cutback coatings for this reason.

Cold-applied asphalt coatings are resistant to deterioration during atmospheric exposure, and so are commonly employed as a heavy-duty coating for aboveground protection where their black color is not objectionable. They do not alligator over prolonged exposure to direct sunlight, but like the coal tars they soften with heat. Asphalt cutbacks are commonly employed in roof coatings.

In formulating asphalt coatings for atmospheric exposure, the asphalt pitch often is upgraded by adding gilsonite—a natural asphalt mined in Utah. Asphalt varnish paints represent an additional upgrading of the material. These are formed by reacting—"cooking"—asphalt with drying oils and resins, and may be pigmented to give dark-colored paints. These are durable, weather resistant, and low in cost, but they show poor gloss retention.

Cleaning solvents and thinners for the coal-tar-base coatings are the aromatic solvents (toluene and xylene) and the aromatic-naphthenic blend coal-tar solvents. Asphalt coating solvents are mineral spirits and painter's naphtha.

Bituminous coatings may be applied by brush or spray to a wire-brushed surface, except for immersion service where a sandblasted surface is desired. Average coverage for the filled, thixotropic, heavy-duty coatings is 60 to 70 sq ft per gal, providing 12- to 15-mils dry film thickness.

Representative spraying conditions for the heavy-duty coatings are:

1. *Air Atomization*
   FLUID PRESSURE: 80 to 100 psig
   ATOMIZING AIR PRESSURE: 70 to 90 psig
   FLUID ORIFICE SIZE: 0.187 to 0.375 in.
   AIR NOZZLE AIR RATE: 16 to 20 cfm at 50 psig
2. *Airless Spray*
   FLUID PRESSURE: 2,500 to 3,000 psig
   FLUID ORIFICE SIZE: 0.021 to 0.026 in.

Catalyzed or Polymerized Coatings

All coatings grouped in this classification undergo a final film cure by means of a chemical reaction—polymerization—which occurs on the surface after application. This results in a coating with superior resistance to the exposure conditions which normally cause film failure.

Solvents are not always employed when the reactive components remain liquid after mixing for a sufficient time to permit application. These "100 percent solids" coatings may be formulated from epoxy, polyester, and polyurethane resins. Without a solvent to evaporate, these 100 percent solids coatings may be applied in pinhole-free, thicker films per coat.

The physical properties of these corrosion-preventive coatings may differ considerably within any generic type, depending upon formulation details. Although representative spraying conditions are presented as guidelines, it is recommended that the manufacturer's specifications always be closely followed. It should be remembered that the manufacturer is just as interested in your receiving satisfactory service from his product as you are.

Catalyzed epoxy coatings: The resins in these coatings consist of a polymerization product of epichlorohydrin and bisphenol A. This resin then undergoes a cross-linking reaction on the metal surface through the action of a basic catalyst, or through the application of heat.

Three types of catalysts are employed, each resulting in epoxy coatings with specific properties:

1. Polyamine catalysts produce hard, high gloss, chemically resistant coatings for exterior atmospheric exposure. These coatings are especially resistant to oil spillage. They are not recommended for immersion service. The disadvantages of polyamine catalysts are the necessity for accurate measurements of small quantities and the toxic nature of the catalyst.

2. Amine adduct catalysts are polyamines partially reacted with an insufficient amount of the epoxy resin. These catalysts produce similar coatings but do not have the disadvantage of need for small quantities, and they are relatively nontoxic.

3. Polyamid catalysts are polyamines reacted with unsaturated fatty acids—usually dimerized linoleic acid. Depending upon the choice of epoxy resin and polyamid employed, epoxy-polyamids range from hard high gloss coatings to 100 percent solids pastes for heavy-duty marine service. These coatings are equal to amine-cured epoxies in atmospheric exposure and are sufficiently superior in resistance to chemicals and water to be suitable for water-immersion service.

The polyamine or polyamid catalyst is added to the epoxy resin just prior to application. This causes a cross-linking of the molecules and curing of the coating after it has been applied to the surface. Because this reaction also will occur in the can of paint before it is applied, there is a limit to the useful life or "pot life" of the coating once the catalyst has been added. The pot life of amine-cured epoxies will vary from 1 to 3 or 4 hr, depending upon temperature. Amine adduct- and polyamid-cured coatings exhibit a considerably longer pot life—as long as 12 to 15 hr at 90°F.

Pot life and film curing time are both increased with decrease in temperature. Below 50 to 60°F, cure time becomes too long to be practical, and forced curing by heat is required. While these coatings dry to touch in a few hours by air drying, several days are required for them to cure completely and develop ultimate chemical resistance.

After stirring in the catalyst, a waiting or induction period of ½ to 1 hr before application is desirable to allow the initial stages of the cross-linking reaction to occur.

Catalyzed epoxy coatings more nearly approach the properties of a baked-on protective coating than do any other cold-cure type of coating. They show excellent resistance to a wide range of chemicals, solvents, oils, acids, and alkalies. In addition, they exhibit excellent adhesion to almost all types of surfaces—metal, wood, and concrete. The polyamid-cured epoxy coatings are exceptional in this respect; the 100 percent solids types will adhere and cure when applied underwater.

They are widely used for coating the interior of crude oil and refined product storage tanks. An excellent coating for water tank interiors results when minor portions of phenolic resins are added in formulating the vehicle. Their resistance to a wide range of corrosive liquids and gases, coupled with good heat-resisting properties, leads to their common use as a heavy-duty coating for industrial plants.

Catalyzed epoxy coatings, like catalyzed phenolics and polyesters, bond well to glass cloth. They are commonly employed in glass cloth laminates or are reinforced with glass cloth.

A suitable cleaning solvent and thinner is a mixture of ketones (MEK or MIBK) and aromatic hydrocarbon solvents (toluene or xylene).

Any practical application method using brush, sprayer, or roller may be employed. Representative coverage for solvent-type coatings is 200 to 250 sq ft per gal per coat, providing 2- to 3-mils dry film thickness. (Solventless pastes can be applied much more thickly per coat.) Brushes and spray equipment used to apply these coatings must be cleaned thoroughly immediately after use. There is no commonly available solvent that will soften and remove the cured epoxy from equipment.

The basic amine catalysts may require special care with respect to health hazards in addition to normal safety precautions. Skin areas contacted by these catalysts should be well rinsed with water. The amine adduct and polyamid catalysts are considerably less toxic.

Representative spraying conditions for the solvent-type coatings are:

1. *Air Atomization*
   FLUID PRESSURE: 20 to 30 psig
   ATOMIZING AIR PRESSURE: 50 to 70 psig
   FLUID ORIFICE SIZE: 0.059 to 0.070 in.
   AIR NOZZLE AIR RATE: 12 to 14 cfm at 50 psig

2. *Airless Spray*
   FLUID PRESSURE: 2,000 to 2,500 psig
   FLUID ORIFICE SIZE: 0.015 to 0.018 in.

Catalyzed phenolic coatings: In coatings of this classification the polymerization reaction between the substituted phenol and the formaldehyde reactants occurs on the metal surface through the action of catalysts (or by heat in the case of baked-on phenolics). Other resins, and sometimes fatty acids, often are included in the reaction to improve film characteristics such as toughness, elasticity, and particularly adhesion to metal surfaces. In this respect, epoxy resins are commonly employed, and phenolic paint primers in particular often are blends of epoxy and phenolic resins.

Catalysts employed usually are polyamines. As with amine-cured epoxies, pot life is short—around 45 min to 1 hr at summer temperatures. Thorough mixing and rapid application are required. Cured phenolic coatings are probably the most solvent resistant of all catalyzed coatings, so thorough cleaning of equipment is required following application.

Catalyzed phenolics are employed as heavy-duty coatings where unusual resistance to acids, alkalies, solvents, and chemicals is required. Strong caustics and strong oxidizing acids will attack these coatings, however. These coatings are one of the very few cold-cured types that will withstand successfully continuous immersion in hot distilled water. They are widely used as linings and patching coatings for process equipment and vessels because of this excellent resistance, even at elevated temperatures.

Phenolics form hard, smooth films with excellent abrasion and wear resistance, and good adhesion to concrete. They therefore are extensively used as floor coatings in chemical plants and food processing industries. They bond very tightly to glass cloth and are commonly employed in glass cloth laminates.

Adhesion to metals is poor, however, so metal primers generally consist of epoxy-phenolic blends. Phenolic resins normally are quite dark in color, and so these coatings usually are not available in white or pastel colors.

A thinner consisting of a 50:50 blend of toluene or xylene with MEK or MIBK is desired. Ketone solvents alone (MEK and MIBK) also will serve.

A sandblasted surface usually is specified with these coatings, since phenolics are commonly applied in immersion service where surface preparation by sandblasting is required. Application may be by brush or spray. Representative coverage is 90 to 100 sq ft per gal per coat, 6 to 8 mils thick when spray applied by the crosshatch technique.

In addition to normal safety precautions, the nature of the catalyst may require special care with respect to health hazards.

Representative spraying conditions are:

1. *Air Atomization*
   FLUID PRESSURE: 20 to 30 psig
   AIR ATOMIZING PRESSURE: 50 to 70 psig
   FLUID ORIFICE SIZE: 0.059 to 0.070 in.
   AIR NOZZLE AIR RATE: 12 to 14 cfm at 50 psig

2. *Airless Spray*
   FLUID PRESSURE: 2,000 to 2,500 psig
   FLUID ORIFICE SIZE: 0.015 to 0.018 in.

Polyurethane coatings: All coatings of this type are formed through the reaction of a polyisocyanate with a substance containing reactive hydrogen—usually a polyalcohol, polyester, or polyether. Because of the large choice of reactants available, films with a wide range of properties may be obtained ranging from hard glossy enamels through soft flexible coatings to foams.

Several types of polyurethanes have been developed, each with specific properties and uses. These are:

1. Two-component polyol-cured formulations. The polyisocyanate fraction usually is partially prepolymerized with a portion of the polyol; otherwise the very short cure time after mixing would make this coating very difficult to apply.

The polyol fraction, usually consisting of derivatives of polyalcohols, polyesters, polyethers, or blends of these, contains the pigment for pigmented coatings. Since presence of moisture causes foaming in polyurethane coatings, this fraction also may contain a small measured amount of water to produce a controlled resilience or "sponginess" in floor coatings.

In polyurethane coatings of this type, a short induction (waiting) period is desired after mixing the two components to permit the cross-linking reaction to start before application.

2. One-package moisture-cured formulations. These coatings consist of solutions of the diisocyanate fraction which have been prepolymerized with as much of the polyol as possible without causing gelling in the can. The final cure is derived from moisture adsorbed from the atmosphere after application.

Because the prepolymer usually is a viscous material, solvents are employed. This is a potentially dangerous formulation since any unreacted toluene diisocyanate (TDI) will escape with the evaporating solvent—so this system is not recommended to be spray-applied.

Moisture-cured polyurethanes generally are not pigmented, as any moisture associated with the pigment would cause the material to gel in the can.

3. Oil-modified formulations. In the usual one-package formulation, the diisocyanate is completely reacted with polyalcohol esters of drying oils. It is thus an oleoresinous paint, and properly should be classified along with alkyds, epoxy esters, and other drying oil paints. Its properties and uses are similar to these oleoresinous paints.

The two-component oil-modified polyurethane is similar to polyol-cured formulations. The one fraction consists of the partially prepolymerized diisocyanate, with the second package containing the remaining required reactant—in this case, polyalcohol esters of fatty acids.

The oil-modified polyurethanes are more flexible, exhibit less yellowing with age, and show better gloss retention than do the other types of polyurethanes. Their chemical resistance is inferior, however, as would be expected from their drying oil content.

When fully cured, polyurethane films exhibit excellent resistance to water and many chemicals and solvents. They are superior in abrasion resistance, and so are becoming widely used as pigmented floor coatings.

The moisture-cured types are commonly employed as floor varnishes where extreme wear resistance is desired, as in gym floors, bowling alleys, etc.

In exterior exposure they retain their flexibility well, but most formulations are subject to serious yellowing and chalking with age. However, oil-modified polyurethanes have been developed which overcome these problems.

Their principal limitation in application is the sensitivity of the isocyanate fraction to moisture. The cans must be kept tightly sealed, especially on humid days. Polyurethanes also exhibit poor adhesion to concrete and steel surfaces unless care is exercised with surface preparation and application.

Suitable solvents are esters such as ethyl or butyl acetate, lacquer thinner, and ketones (MEK and MIBK).

Surface preparation of metals by sandblasting is preferred. Any common method of application is suitable when applying polyurethane paints and coatings. Heavy-duty floor coatings usually are applied by pouring and troweling or by a curtain coater. Representative coverage is 250 to 300 sq ft per gal per coat, providing 2-mils dry film thickness.

The diisocyanate portion must be recognized as a health hazard, primarily due to the presence of toluene diisocyanate (TDI). This chemical is very toxic, causing severe eye and skin irritation when contacted in low concentrations. Appropriate care must be taken during spray application in addition to normal safety precautions by wearing air-supplied hoods or respirators. This is especially true of moisture-cured formulations.

Representative spraying conditions for the one-package moisture-cured clear varnish are:

1. *Air Atomization*
   FLUID PRESSURE: 10 to 25 psig
   ATOMIZING AIR PRESSURE: 30 to 50 psig
   FLUID ORIFICE SIZE: 0.040 to 0.052 in.
   AIR NOZZLE AIR RATE: 12 to 14 cfm at 50 psig
2. *Airless Spray*
   FLUID PRESSURE: 1,800 to 2,000 psig
   FLUID ORIFICE SIZE: 0.011 to 0.013 in.

Catalyzed bitumen-epoxy coatings: Two types of coatings have been developed which fall in this classification: the well-known coal-tar epoxies and the newer asphalt epoxies. Other combinations have been developed such as the coal-tar urethanes, but since these are not yet in general use they will not be discussed further.

The coal-tar epoxy and asphalt epoxy coatings are blends of selected coal-tar pitches and asphalt pitches with epoxy resins, and are catalytically cured using polyamine and polyamid catalysts. The polyamids produce more flexible coatings and show a longer pot life than do the polyamine-catalyzed coatings, which have a pot life of 1 to 4 hr depending upon temperature.

The heavy-bodied nature of the bitumen epoxy means that special care must be taken in adding the catalyst. Hand stirring is inadequate; power stirrers must be used. After application, these coatings must be recoated within 6 to 24 hr, depending upon temperature, or delamination between coats may occur.

The bitumen-epoxy coatings combine the excellent properties of both the epoxy and the bituminous coatings. In addition, the presence of coal-tar or asphalt pitches improves the water resistance of the epoxies and decreases their cost, while the epoxy resin eliminates high temperature softening and alligatoring of the coal-tar coatings and greatly improves their solvent resistance.

These coatings are designed for heavy-duty service. They show excellent adhesion to metal and nonmetal surfaces, very good abrasion and weathering resistance, and superior resistance to acids, alkalies, chemicals, oils, water, and solvents (except for the aromatic solvents toluene and xylene). They provide long-lasting protection in immersion service in petroleum oils and industrial waters and brines. The coal-tar epoxies cannot be used in drinking-water service because of the phenol odor and taste imparted to the water. Asphalt epoxies do not have this limitation, however.

Their high solids content (70 to 90 percent by volume) allows thick films to be applied per coat. Two coats of 8 to 10 mils each are recommended for most services. In addition, they are suitable for reinforcement by glass cloth when desired for edge protection, patching in tanks, etc.

The main uses of bitumen epoxy coatings are for the protection of dam gates, piers, and other equipment such as barge and ship bottoms exposed to or immersed in fresh or salt water. The exterior surfaces of buried structural steel and pipe and the interior bottoms of crude storage tanks are commonly protected by coal-tar epoxies. In addition, their high degree of impermeability to gases, particularly hydrogen sulfide, also makes them well adapted for the protection of the vapor zones of sour crude tanks.

The catalyzed coating must be applied within a few hours after the addition of catalyst, otherwise it will harden in the can. Brushes and spray equipment used to apply these coatings must be cleaned thoroughly immediately after use. There is no commonly available solvent that will soften and remove the cured material from equipment. Suitable cleaning solvents and thinners are toluene, xylene, and preferably a blend of these with ketones (MEK or MIBK).

These coatings may be applied by brushing, troweling, or spraying. The thick, heavy-bodied consistency of most of these formulations makes spraying difficult without properly designed equipment. Some thinning with solvent (1 to 1½ pints per gal) is required for spraying.

A well-scraped and wire-brushed surface is sufficient for most services except immersion service, where sandblasting is required. Representative coverage is 100 to 150 sq ft per gal per coat, providing 6- to 8-mils dry film thickness.

The polyamine and polyamid catalysts will require special care with respect to health hazards in addition to normal safety precautions. Skin areas contacted by these catalysts should be well rinsed with water.

Representative spraying conditions are:

1. *Air Atomization*
   FLUID PRESSURE: 40 to 50 psig
   ATOMIZING AIR PRESSURE: 60 to 80 psig
   FLUID ORIFICE SIZE: 0.110 to 0.187 in.
   AIR NOZZLE AIR RATE: 16 cfm at 50 psig
2. *Airless Spray*
   FLUID PRESSURE: 2,000 to 2,500 psig
   FLUID ORIFICE SIZE: 0.018 to 0.021 in.

Chlorosulfonated polyethylene (Hypalon) coatings: Chlorosulfonated polyethylene (Hypalon), a development of E. I. Du Pont de Nemours & Co., is prepared by treating polyethylene with chlorine and sulfur dioxide. At the time of application, the resulting product is further reacted by cross-linking (vulcanization) into a rubbery coating material. To increase film density and hardness, phenolic and other resins are sometimes incorporated in the coating formulations.

Curing takes place with the aid of a lead-containing alkaline catalyst, and the catalyzed material has a limited pot life. The film after cure is characterized by its great flexibility.

Chlorosulfonated polyethylene paints are suitable for use under severely corrosive conditions. They are very resistant to the halogens (chlorine, bromine, etc.). They are one of the few types of coatings to resist successfully the action of hot water. They show excellent resistance to the deteriorating effects of oxygen, ozone, and sunlight, and so provide excellent service in atmospheric exposure, especially when pigmented. Thinners and cleaners are toluene or xylene, preferably with additions of a ketone (MEK or MIBK).

Hypalon paints may be applied by roller, brush, or spray. Although they exhibit fair to good adhesion to surfaces, chlorinated rubber primers usually are employed when application is made to metal surfaces. Surface preparation must be by sandblasting to clean metal when resistance to corrosive conditions is desired. Representative coverage is 50 to 100 sq ft per gal per coat, providing 3- to 5-mils dry film thickness.

The lead-containing catalyst present in certain formulations of these paints may require special care with respect to health hazards in addition to normal safety precautions. Any catalyst contacting the skin should be washed off immediately.

Representative spraying conditions are:

1. *Air Atomization*
   FLUID PRESSURE: 10 to 20 psig
   ATOMIZING AIR PRESSURE: 50 to 70 psig
   FLUID ORIFICE SIZE: 0.046 to 0.070 in.
   AIR NOZZLE AIR RATE: 12 to 14 cfm at 50 psig
2. *Airless Spray*
   FLUID PRESSURE: 2,000 to 2,500 psig
   FLUID ORIFICE SIZE: 0.013 to 0.015 in.

Polyester paints and coatings: The polyester resin is a condensation product of an organic acid and an unsaturated polybasic alcohol. The resulting unsaturated polyester resin is further reacted (at the time of application) by cross-linking with styrene through the action of the MEK peroxide catalyst and an accelerator (cobalt naphthenate).

Since the polyester resin dissolves in the styrene monomer to form a stable fluid solution, polyester coatings usually are of the 100 percent solids type. Once the peroxide catalyst and accelerator (promoter) are added, the cross-linking reaction is very rapid and gelling occurs in a matter of minutes. Consequently, specialized application equipment is required which meters and mixes catalyst, accelerator, and resin immediately before application to the surface (see page 176).

Polyester coatings usually are the 100 percent solids type with no solvent to evaporate, so they may be applied in thick films with no pinholing. They exhibit high gloss with good resistance to weathering. They show excellent acid and solvent resistance, but poor resistance to action of alkalies. They combine readily with silicone resins to form high gloss porcelainlike paints for decorative and heat-resistant finishes on metal products.

The principal shortcomings of polyester resins are their brittleness and the shrinkage that occurs during cure. Although some formulations containing modifying resins have overcome these problems, polyesters more commonly are reinforced with glass or asbestos fibers to which the resin bonds very tightly.

Polyester coatings usually are applied through specialized spray equipment. In one system, the metered MEK peroxide catalyst is injected in the compressed air used for atomizing the polyester resin. In another system, polyester resin containing the accelerator is atomized through one spray gun head, meeting the atomized resin with catalyst as it streams from the second head of a two-headed gun. Airless spray application also is accomplished by a two-headed gun atomizing the resin with accelerator and resin with catalyst through separate heads. In another system, resin with accelerator is atomized through both heads, with the metered catalyst being injected into the center of the converging streams. Chopped glass fibers often are injected into the atomized polyester stream by means of fiber glass choppers attached to the gun.

Polyester resins with slower gel times are commonly applied by brushing or rolling onto glass reinforcing mats, saturating the mat and cementing it to the surface to which it is applied. Suitable cleaning solvents, and thinners when desired, are the ketones (MEK and MIBK).

Special safety precautions must be observed in the application of polyester coatings. Only stainless steel and aluminum parts may be permitted to contact the MEK peroxide catalyst or resin containing the catalyst, otherwise explosive decomposition of the peroxide may occur. Consequently, only pumps and guns conforming to this metallurgical limitation can be used. Furthermore, the volatile styrene monomer is toxic, and the applicator must be protected from its inhalation through the use of air-supplied hoods or respirators.

Representative spraying conditions presented below reflect the broad range of viscosities in polyester coatings. In air atomization, only the minimum air pressure required to atomize the fluid should be employed. For gel-coat lay-ups, fluid orifice sizes larger than 0.110 in. often are employed.

1. *Air Atomization*
   FLUID PRESSURE: 40 to 70 psig
   AIR ATOMIZING PRESSURE: 20 to 40 psig
   FLUID ORIFICE SIZE: 0.086 to 0.110 in.
   AIR NOZZLE AIR RATE: 14 to 16 cfm at 50 psig

2. *Airless Spray*
   FLUID PRESSURE: 1,200 to 1,500 psig
   FLUID ORIFICE SIZE: 0.015 to 0.021 in.

Zinc-rich primers and coatings: Zinc-rich coatings are finding ever-increasing application as a heavy-duty corrosion-resistant primer, because they essentially eliminate one of the principal causes of protective coating failure: film lifting by undercutting corrosion on the metal surface. Zinc-rich primers contain a sufficiently high percentage of metallic zinc dust to be electrically conductive, and so serve to provide galvanic protection to the metal surface in much the same manner as is provided by hot-dip zinc galvanizing.

The zinc dust may be suspended in silicate solutions which form an inorganic zinc silicate binder on the metal surface, or suitable organic coating resins may be employed to form organic binders for the zinc dust. The zinc silicate primers will be discussed first.

Two types of zinc silicates have been developed and are widely used: the original postcure inorganic zinc silicate and the self-cure zinc ethyl silicate.

1. The postcure zinc silicate primer is a three-package system, consisting of a water solution of sodium, potassium, or lithium glassy silicates; a can of finely divided zinc dust (sometimes with a minor portion of lead dust); and a container of the acidic curing solution. The zinc dust is added to the silicate solution with stirring, after which it is strained and allowed to stand (with occasional stirring) for ½ to 1 hr before applying. When the applied film is dry (indicated by a color change from pink to gray), the acidic curing solution is liberally applied. This zinc silicate primer may be topcoated after 5 to 7 days cure time, but only after the catalyst residue is first washed from the surface. If the topcoat is a vinyl, a vinyl primer must be applied first for adhesion.

2. The self-cure zinc silicate primer usually is a two-package system, consisting of an alcohol or cellosolve ester solution of silicate esters such as tetraethyl silicate, with a separate container of the zinc dust. The zinc dust is added to the silicate solution and applied in the same manner as with the postcure type. The self-cure zinc silicate film is not subject to water damage after the film has dried thoroughly (1 to 2 hr)—in fact, after the initial drying period, wetting with water will aid in its cure. After 1 to 2 days cure time, topcoats may be applied without the necessity of washing off the zinc silicate surface.

The cured inorganic zinc silicate coating consists of metallic zinc dust dispersed in a zinc silicate binder, and in the simplest terms can be considered a cross between galvanizing and a ceramic. The inorganic matrix is a conductor of electricity, and permits sacrifice of some of the zinc metal contained in the coating to cathodically protect the base metal from corrosion due to small "holidays" which result from misapplication, or from damaged areas in the coating. To provide this cathodic protection the coating must contain at least 80 percent by weight of zinc—most zinc silicates approach 90 percent zinc dust. Only one coat is applied to give a dry film thickness of from 2 to 5 mils. Heavier coatings will mudcrack and must be removed and reapplied. To provide cathodic protection (as is normally desired when applying this type of primer), application must be to a sandblasted surface and preferably to a white finish for immersion service.

During the initial curing of the film, zinc silicate films are damaged by any moisture or water contacting the surface. Therefore, in humid periods coating application should be scheduled for morning or early afternoon hours. Water contacting the surface after this initial drying period will do no harm, but instead may actually aid in the final cure of the film.

These coatings are characterized by their outstanding resistance to immersion in oils, hydrocarbons, and organic solvents of all types. They also are outstanding in resistance to abrasion and to deterioration from atmospheric exposure. However, since they contain metallic zinc, they are limited in their exposure to acids and caustics.

Zinc silicate primers applied for exposure to or immersion in aqueous solutions, fresh or salt water, or marine atmospheres require a protective topcoat for lasting service. The topcoat is selected which is suitable for the exposure condition. Common seal coats are epoxy esters and catalyzed epoxies, vinyls, and chlorinated rubber. The coal-tar and asphalt epoxies provide excellent service in marine exposures as heavy-duty topcoatings.

They are used to protect stationary and mobile marine equipment and to protect interior surfaces of petroleum tankage and clean oil tankers. They also may be used as a permanent corrosion-preventive prime coat for chemically resistant coatings in chemical exposure or exposure to weathering under humid or corrosive conditions.

Zinc silicate coatings successfully resist deterioration at elevated temperatures—up to 800 to 1,000°F. Consequently, they are excellent as primers under silicone stack paints, etc.

Organic zinc-rich primers consist of zinc dust (85 to 93 percent by weight) carried in a suitable organic coating vehicle. Experience has shown that the most satisfactory binders are the polyamid-cured epoxy resins, chlorinated rubbers, and polyesters formulated with polyethers. These may be packaged with the zinc dust in a separate container or already dispersed in the vehicle. In catalytically cured types the catalyst fraction is packaged separately.

Organic zinc-rich primers are similar to the zinc silicates by providing superior protection through galvanic action. They are more flexible, less sensitive to moisture during cure, and less demanding of surface preparation quality than are the zinc silicates. Abrasion resistance and durability are about equal, but resistance to organic solvents and oils and to high temperatures is considerably less than with the inorganic silicates, as might be expected.

The following discussion on solvents and application conditions refers only to the zinc silicates. The organic zinc-rich primers are applied in a manner similar to coatings having the same vehicle.

No thinners should be used with the inorganic zinc silicates. Spilled material may be removed with water. Isopropyl alcohol aids in cleaning equipment used with the self-cure zinc silicate primers.

Zinc silicate coatings may be applied using conventional pressure-spray equipment with minor modifications discussed below. Spray application is preferred to brushing. These coatings require a very good surface preparation. A commercial blast finish is considered the minimum acceptable surface preparation.

The pressure feed of zinc silicates is handled most successfully by pressure tanks equipped with a bottom outlet and a stirrer. Constant stirring is important. It is also important to locate the pot at about the same elevation as the job. Zinc silicates are very heavy, and too great a pressure drop occurs if they are lifted by pot pressure more than a few feet.

All coatings of this type are applied to a thickness of 2 to 4 mils in one coat only. The minimum and maximum thickness limits of 2 mils and 5 mils should be maintained. Two mils is desired for high temperature service. Mudcracking may occur with coats thicker than 5 mils, and the coating must be removed completely.

Average coverage is 200 to 250 sq ft per gal for the one coat applied at 3- to 4-mils thickness.

Representative spraying conditions are:

1. *Air Atomization**

   FLUID PRESSURE: 15 to 25 psig

   ATOMIZING AIR PRESSURE: 40 to 60 psig

   FLUID ORIFICE SIZE: 0.070 to 0.086 in.

   AIR NOZZLE AIR RATE: 15 to 18 cfm at 50 psig

2. *Airless Spray*

   The application of zinc silicates by airless spray generally is not recommended because of buildup of zinc in the pump valves and packing and around the gun nozzle. However, certain formulations have been applied sucessfully under one or the other of these two representative airless spray conditions:

   FLUID PRESSURE: 2,200 to 2,500 psig

   FLUID ORIFICE SIZE: 0.018 to 0.021 in.

* *Note:* A nylon fluid needle must be used, with leather packings. Also, as with all water-base paints, stainless steel nozzle setups are recommended.

or

FLUID PRESSURE: 800 to 1,200 psig
FLUID ORIFICE SIZE: 0.026 to 0.031 in.

Water-base Paints and Coatings

Water-base (water-dispersion) paints first were developed as interior house paints for plaster and gyp-board surfaces, employing binders of styrene or styrene-butadiene latexes. The term "latex paints," which was proper with these, is now being applied incorrectly to all water-base paints.

Because of similar properties and applications, the different water-base paints will be discussed together. These fall into two broad groups according to usage: the interior and exterior house paints for cementiferous surfaces, and the newer protective coatings for metals. (The specialized coal-tar dispersion coatings are mentioned in the discussion, "Bituminous coatings (cold-applied)," and water-base zinc silicates are reviewed in the discussion, "Zinc-rich Primers and Coatings.")

The interest in developing improved water-base paints and coatings has increased greatly in recent years since the introduction of smog control legislation. This legislation stringently limits the use of photochemically reactive solvents, which, unfortunately, are the more commonly employed aromatics and ketones. (See Chap. 3, "Health and Safety Considerations," for additional information on this subject.)

Water-base house paints are dispersions of polystyrene and styrene-butadiene copolymer latexes, and of the widely used polyvinyl acetate and polyacrylic resins and copolymers of these. They all exhibit excellent adhesion to plaster, cement, and wood surfaces, and have been widely accepted as interior and exterior house paints because of their easy application and cleanup, fast drying, low odor, and good hiding power. Other obvious advantages are the nontoxic and nonflammable properties of the water solvent.

Since water-base paint films are porous, and the resins are resistant to deterioration by alkali salts, they can be applied safely to plaster and cement which is not completely cured or dry. One recommended application is as a barrier or seal coat on cement or plaster which is to be finished with an oleoresinous paint.

The water-base paints are employed for atmospheric exposure only. They are not available in glossy finishes, although developments of water-base enamels are being forecast by manufacturers. They show relatively short life on exterior wood surfaces as compared with the oleoresinous paints. And as might be expected, cans of the paint must be protected from freezing during storage.

The latex-type paints show a limited shelf life as compared with the other water-base paints. The stability of the latex is dependent upon maintaining the proper basicity through use of ammonia or substituted amines. The polyacrylic water-dispersion paints, on the other hand, are exceptionally stable during storage.

Corrosion-protective coatings of water-dispersed resins have been developed based on polyacrylics, alkyd-acrylic copolymers, and polyvinyl acetate acrylic copolymers. Primers with corrosion-inhibitive pigments such as red lead also are available. These dispersions usually contain nonvolatile resinous oils to seal the porosity normally present in the film.

Water-base protective coatings are reported to provide excellent service on steels in industrial atmospheres, with good resistance to moisture, oil spillage, and acid and alkali fumes. They are fast-dry materials, permitting the application of several coats in a day. They are not suitable for immersion service, show limited resistance to solvent fumes and spillage, and, of course, must be applied to oil and grease-free surfaces.

Thinning generally is not recommended with water-base coatings. Water is used for equipment cleaning; additions of MEK help in removing partially dried material. Do not shake water-base paints—stir only. Shaking will emulsify air throughout the material, and this will take hours to separate out again.

Methods of application are by roller, brush, or spray. The interior wall paints commonly are applied by roller-coating, while exterior surfaces are sprayed. Airless spray application is not recommended with the latex types because the high pressure pump action tends to break down the dispersion. Average coverage per gallon is 250 to 300 sq ft per coat, giving a 1½- to 2-mils dry film thickness.

Representative spraying conditions are:

1. *Air Atomization**
   FLUID PRESSURE: 15 to 30 psig
   ATOMIZING AIR PRESSURE: 30 to 45 psig
   FLUID ORIFICE SIZE: 0.070 to 0.086 in.
   AIR NOZZLE AIR RATE: 14 to 15 cfm at 50 psig

2. *Airless Spray*†
   FLUID PRESSURE: 1,800 to 2,000 psig
   FLUID ORIFICE SIZE: 0.015 to 0.018 in.

* *Note:* As with all water-base paints, stainless steel nozzle setups are recommended.

† *Note:* A preorifice insert aids in the airless atomization of polyvinyl acetate paint.

Miscellaneous Paints

Three paint systems are discussed in this section which are formulated for highly specialized services or purposes. These are:

- Fire-retardant paints, which reduce the spreading of a fire through their foaming action under heat
- Fluorescent paints, which seem to glow due to the conversion of invisible ultraviolet light into light of visible wavelengths
- Antifouling paints, applied to marine structures to reduce the growth of marine organisms on submerged surfaces

Fire-retardant paints: Fire-retardant paints generally are formulated with nonflammable binders, and with pigments that decompose under heat to give off a smothering gas, at the same time causing the paint film to foam.

Vehicles employed are those which will not support combustion, such as those formulated with silicones, chlorinated rubbers, chlorinated paraffin, and polyvinyl chloride resins. Film upgrading often is accomplished by adding alkyd resins, in particular to the chlorinated paraffins. Water-base fire-retardant paints also are available.

Pigments are selected which decompose upon heating to give off a smothering gas—carbon dioxide, nitrogen, or chlorine. These include zinc and antimony oxides, calcium and lead carbonates, zinc borate, and lead and magnesium silicates. Inactive colored pigments sometimes are added for decorative effects.

A high pigment/vehicle ratio is employed for effective action, so the most effective types of fire-retardant paints are flat or low gloss coatings. Fire-retardant enamels also are available when desired for decorative effects. To obtain adequate fire-protection activity, these paints should be applied in sufficiently thick films. Dry paint films of from 5 to 8 mils generally are recommended. When heated, a fire-retardant paint film will expand in thickness by foaming, to seal the substrate and act as a thermal insulator. This will slow down the time for the surface on the opposite side of the wall to reach combustion temperatures.

Fire-retardant paints, to be effective, cannot be applied over old conventional paints since these will blister and spall off, carrying the fire-retardant paint with them.

Suitable solvents and thinners will depend upon the vehicle of the paint. Since commonly employed vehicles are chlorinated rubber or alkyds containing chlorinated compounds, aromatic solvents (toluene and xylene) would be a good first guess as a cleaning solvent—unless, of course, the proper solvent is given.

No representative spraying conditions may be given because of the widely differing characteristics of fire-retardant paints.

Fluorescent paints: Fluorescent paints have the property of absorbing invisible ultraviolet light, converting it into light of visible wavelengths and emitting it in the same colors as its reflected light. These paints "give off" several times more light in their color range than would be reflected from ordinary paints of the same color.

Pigments in these paints consist of highly fluorescent dyes dissolved in special clear colorless thermoplastic resins, which are then ground to pigment size. A high pigment loading of around 50 percent is used in the paints so that an effective amount of fluorescent dye will be present in the applied film.

The vehicle in fluorescent paints is chosen for its weatherability, resistance to moisture, and water-white clarity which will not darken with age. Polyacrylic lacquers are commonly employed.

Fluorescent paints have been developed which are lightfast and effective for about one year's exposure in bright sunlight. The fluorescent dyes range in color from yellow to red, but inert dyes of other colors may be added. Even with a 50 percent pigment loading these paints have low hiding power, so they are always applied over a flat or semigloss white undercoat which enhances the fluorescence due to its reflectivity.

A total dry film thickness of at least 3 mils is desired. For exterior exposure, effective life is considerably increased by topcoating with a special ultraviolet light-absorbing clear polyacrylic lacquer.

Fluorescent paints are very effective when used for warning signs and markings, traffic stripes, etc. Due to their fluorescent properties they will maintain their colored appearance under dim light when other paint colors will appear gray.

Thinning and cleaning solvents are the aromatic hydrocarbons toluene and xylene. Ketones should not be used since they may dissolve out some of the dyes.

Fluorescent paints usually are applied by brushing or spraying. Representative coverage is 300 to 400 sq ft per gal per coat, providing 1- to 1½-mils dry film thickness. A total dry film thickness of at least 3 mils must be applied for lasting effectiveness.

Representative spraying conditions are:

1. *Air Atomization*

   FLUID PRESSURE: 10 to 15 psig
   ATOMIZING AIR PRESSURE: 40 to 50 psig
   FLUID ORIFICE SIZE: 0.040 to 0.052 in.
   AIR NOZZLE AIR RATE: 12 to 14 cfm at 50 psig

2. *Airless Spray*
   FLUID PRESSURE: 1,300 to 1,600 psig
   FLUID ORIFICE SIZE: 0.011 to 0.013 in.

Antifouling paints: Antifouling paints are applied as the topcoat to submerged surfaces of marine structures—particularly ships—in order to reduce the rate of growth of barnacles and other marine organisms. The action of these paints is due to the slow leaching of poisonous pigments to the surface of the film.

Successful antifouling paints depend upon the careful selection of poisonous pigments and the proper choice of binders having limited life. Pigments with slow solubility in seawater and which are highly poisonous to the marine organisms are required. Experience has shown that the most effective are cuprous oxide, mercuric oxide, and mercurous chloride, with cuprous oxide being the one most widely employed. To be effective over the required life of the coating, from 30 to 40 percent by weight of these pigments are employed. Additional inactive pigment extenders such as iron oxide often are included for a proper pigment/vehicle ratio.

The binders for these films must be formulated to undergo a slow film failure through erosion. A useful life of only 9 to 12 months is desired, since most ships drydock with that frequency. Nearly all binders are blends of rosin with synthetic resins, such as esterified rosin (ester gum), chlorinated rubber, phenolics, and vinyls. Most binders providing a 1-year life contain about 50 percent rosin. Longer-lasting antifouling paints will result from use of the synthetic resin alone, but to be effective these paints must contain a greater loading of cuprous oxide—up to 80 percent.

Antifouling paints must be applied over a long-lasting high quality anticorrosive primer. The primer should outlast several recoatings of antifouling paint. The two functions of corrosion protection and antifouling cannot be combined in one coating material, however, since the copper salts would greatly accelerate pitting of the steel hull if contacted by them. Effective marine primers which may be topcoated satisfactorily with antifouling paints are coal-tar epoxies, zinc silicates, and organic zinc-rich primers—especially those based upon the epoxy polyamid resins.

Antifouling paints must be formulated to be fast drying and easy to apply, as drydocking charges are high. Any commonly employed application method—roller, brush, or spray—is suitable for their application. Frequent stirring during application is required because of the dense nature of their copper and mercury pigments.

Since various types of paint vehicle may be employed, no recommendations as to cleaning solvents or spraying conditions can be given. If the proper thinning and cleaning solvent is not known, mineral spirits could be tried.

Tabular Presentation of Paint and Coating Data

The essential data in the foregoing discussion on paints and coatings are repeated in tabular form in Tables 1-1 and 1-2.

The physical properties that govern coverage, film thickness, drying time, etc., may vary considerably within any one generic type of paint. This is especially true of the protective coatings. Consequently, the data presented in these tables should be used only as a quick comparative reference for checking the characteristics of different types of paints and coatings.

Paint Pigments

Pigments extend paint life by hardening and toughening the film, by neutralizing acids formed through decomposition of the binder, and by absorbing and reflecting ultraviolet rays which otherwise would deteriorate the resin binder. Consequently, paints are more weather-resistant than are clear unpigmented varnishes.

Several classes of pigments have been developed to serve specific purposes:

1. *Nonreactive white pigments:* for hiding power and whiteness
2. *Reactive white pigments:* for acid neutralization and corrosion inhibition, as well as for hiding power and whiteness
3. *Extenders:* to enhance the properties of pigments, to reduce film cracking, and to reduce cost
4. *Corrosion inhibitive pigments:* to reduce corrosion of metal substrates by reacting with the initial corrosion product to form insoluble metal salts
5. *Colored pigments:* to add color to the paint film.

The hiding power, or "opacity," of a pigment depends upon the degree to which the refractive index of the pigment is greater than that of the binder, which for oleoresinous vehicles is around 1.5. Hiding power is expressed as the square feet of a standard black-and-white surface that will be hidden per pound of pigment when dispersed in a linseed oil vehicle.

Tint strength is determined as being proportional to the amount of ultramarine blue or carbon black pigment required to yield a standard brightness when mixed in a linseed oil paste of the white pigment to give a total pigment concentration of 46 percent by volume (46 percent

TABLE 1-1 Application Characteristics of Generic Types of Paints and Coatings

Generic type	Resin composition	Drying or curing mechanism	Solvents and thinners	Coverage/coat sq ft/gal	Dry time to recoat, hrs	Dry film thickness per coat, mils	Total dry film thickness: Atmospheric exposure film, mils	Total dry film thickness: Immersion film, mils
Drying oil	Unsaturated drying oil and resin	Solvent evaporation, oxygen adsorption	Mineral spirits; turpentine	400–450	30–40	2–2½	4–5	
Alkyd	Polyacid, polyalcohol, drying oil, coreacted	Solvent evaporation, oxygen adsorption	Mineral spirits; turpentine	350–400	15–25	1–1½	4–5	
Phenolic varnish	Substituted phenol, formaldehyde, drying oil, coreacted	Solvent evaporation, oxygen adsorption	Mineral spirits and aromatics	250–350	10–15	1½–2	5–6	8–10
Epoxy ester	Epichlorohydrin, bisphenol A, drying oil, coreacted	Solvent evaporation, oxygen adsorption	Mineral spirits and aromatics	350–400	20–30	1–1½	4–5	
Chlorinated rubber	Chlorinated natural rubber	Solvent evaporation	Aromatics	300–350	10–15	1–1½	5–6	8–10
Vinyls	Polyvinyl chloride, polyvinyl acetate copolymerized	Solvent evaporation	Ketones	200–250	1–2	¾–1	3–4	6–8
Silicones	Modified silicone	Solvent evaporation	Aromatics	350–400	5–10	¾–1	1½–2	
Acrylic lacquer	Modified polyacrylic	Solvent evaporation	Ketones; lacquer thinner	350–400	1–2	¾–1	2–3	
Bituminous (cold-applied)	Coal-tar or asphalt pitch	Solvent evaporation	Aromatics; mineral spirits	60–70	40–50	12–15	20–25*	50–75
Catalyzed epoxy	Epichlorohydrin, bisphenol A, copolymerized	Polyamine or polyamid catalyst	Ketones-aromatics mix	200–250	5–10	2–3	5–6	10–12
Catalyzed phenolic	Substituted phenol, formaldehyde, copolymerized	Polyamine or acid catalyst	Ketones-aromatics mix	100–150	10–15	4–6	4–5	12–15
Moisture-cured polyurethane†	Polyisocyanate, polyol, copolymerized	Polymerization	Esters; ketones	300–400	3–5	2–3	5–6	
Bitumen epoxy	Asphalt or coal-tar pitch, epoxy resin	Polyamine or polyamid catalyst	Ketones-aromatics mix	100–150	10–15	6–8	10–12	15–20
Hypalon	Chlorosulfonated polyethylene	Lead-containing alkali catalyst	Ketones-aromatics mix	50–100	15–25	3–5	4–6	10–12
Polyester	Organic acid, polyalcohol, styrene, copolymerized	Polymerization	Ketones	150–300	½–1	5–10	5–10	15–20
Organic zinc-rich	Zinc dust in epoxy or chlorinated rubber	Solvent evaporation, or catalyst (epoxy)	Aromatics	150–250	10–15	2–4	2–4	2–4‡
Zinc silicate	Zinc dust in alkali silicates	Polymerization	Water	200–250	25–40§	3–5	3–5	3–5‡
Water base	Styrene-butadiene, PVA, or polyacrylic, in water	Water evaporation	Water	250–350	5–10	1½–2	2–4	
Fire retardant	Nonflammable resin	Solvent evaporation	Depends on resin base	200–250	5–10	2–3	6–8	
Fluorescent	Polyacrylic	Solvent evaporation	Ketones; lacquer thinner	300–400	1–2	1–1½	3–4	
Antifouling	Rosin, resin, coreacted	Solvent evaporation	Aromatics or mineral spirits	250–300	5–10	2–3	....	4–6

Notes:
* In atmospheric service, coal-tar coatings must be overcoated with a coal-tar emulsion.
† Characteristics of other types of polyurethanes are presented in the preceding discussion.
‡ Zinc-rich primers usually are top-coated with a suitable coating for immersion service.
§ Postcure zinc silicates require 3- to 5-days cure time before top coating.

TABLE 1-2 Service Recommendations of Generic Types of Paints and Coatings

Generic type	Outstanding properties	Suitable substrate	Recommended service conditions		Services not recommended
			Atmospheric exposure	Immersion service	
Drying oil	Surface wetting; adherence; weatherability	Wood; steel	Uncontaminated atmospheres	Not used	Industrial atmospheres
Alkyd	Gloss retention; durability; weatherability	Wood; steel	Industrial and humid atmospheres	Not used	Chemical fumes
Phenolic varnish	Moisture resistance; weatherability	Concrete; wood; steel	Humid and marine atmospheres	Intermittent service, potable water	Chemicals; acids; alkalies
Epoxy ester	Durability in humid and chemical atmospheres	Wood; steel	Industrial and chemical atmospheres	Not used	Acids; alkalies
Chlorinated rubber	Adhesion; alkali and water resistance	Concrete; wood; steel	Corrosive and chemical atmospheres	Acids; alkalies; salts; water	Strong solvents and acids
Vinyls	Durability; chemicals and water resistance	Concrete; steel	Corrosive and chemical atmospheres	Acids; alkalies; salts; water	General maintenance
Silicones	High temperature resistance	Steel	Hot surfaces (up to 1000°F)	Not used	All other services
Acrylic lacquer	Gloss retention; weatherability	Wood; steel	Uncontaminated atmospheres	Not used	Industrial atmospheres
Bituminous (cold-applied)	Asphalt: Heavy-duty atmosphere Coal tar: Water immersion	Concrete Steel	Heavy-duty black (asphalt)	Water and brine: Heavy duty (coal tar)	Oils and solvents; colors
Catalyzed epoxy	Adhesion; chemical, acid, alkali resistance	Concrete; wood; steel	Humid and chemical atmospheres	Water; salts	General maintenance
Catalyzed phenolic	Abrasion resistance; chemical, acid, hot-water resistance	Concrete; steel	Humid and chemical atmospheres	Hot, cold water; salt solutions	Alkali immersion
Moisture-cured polyurethane	Wear resistance	Wood; steel	Heavy-duty floor varnish	Not used	Exterior exposure
Bitumen epoxy	Heavy-duty atmosphere and water immersion	Concrete; steel	Heavy-duty industrial and marine	Water; sour oil	Potable water immersion
Hypalon	Flexibility; weatherability; hot-water resistance	Concrete; steel	Heavy-duty chemical and marine	Hot, cold water; salts; chemicals	General maintenance
Polyester	Acid and solvent resistance	Concrete; steel	Humid atmospheres	Water; acids; crude oil	Alkali immersion
Organic zinc-rich	Heavy-duty metal primer	Steel	Humid and industrial atmospheres	(Usually top coated)	Immersion without top coat
Zinc silicate	Heat and abrasion resistance; heavy-duty metal primer	Steel	Humid and industrial atmospheres	Oils; solvents (without top coat)	Acids; alkalies (without top coat)
Water base	Good hiding; no odor; nonflammable	Wood; concrete; plaster	Interior and exterior on concrete; plaster; wood*	Not used	Industrial atmospheres
Fire retardant	Nonflammable; foam with heat	Wood; concrete; plaster	Interior surfaces for fire protection	Not used	Exterior exposure
Fluorescent	Fluoresces under light	Wood; steel	Uncontaminated atmospheres	Not used	Industrial atmospheres
Antifouling	Toxic to marine organisms	Steel	Not used	Brackish and sea water	Any other service

* Recommended only for interior exposure.

PVC). It is an arbitrary value, related to white lead with a given value of 100.

The refractive index, hiding power, and tint strength of a number of pigments are presented in Table 1-3, where it is seen that rutile titanium dioxide has the highest refractive index of all pigments.

The hiding power of a pigment depends upon the pigment volume concentration (PVC) in the binder. As the PVC increases, the hiding power per pound of pigment decreases due to crowding of the particles and consequent reduction of light scattering. This is illustrated in Fig. 1-1a and b for titanium pigments in an alkyd vehicle.

1. Nonreactive white pigments: In modern paints these pigments now consist solely of the rutile and anatase crystal forms of titanium dioxide, because of their great hiding power and tint strength (see Table 1-3). The rutile form is a nonchalking pigment—a characteristic desired to a considerable degree in most exterior paints to increase weather resistance. Anatase titanium dioxide is added for controlled film chalking in chalking white paints and in other exterior paints where some chalking is desired as a method of maintaining a clean and white surface.

Less effective nonreactive white pigments are zinc sulfide and lithopone (zinc sulfide-barium sulfide coprecipitated), and these are little used in modern paints. They are less resistant to weathering and are considerably lower in hiding power than is titanium dioxide.

2. Reactive white pigments: These pigments contribute to hiding power, but their main purpose is to neutralize acids that form in oleoresinous films during exposure, thereby extending paint film life.

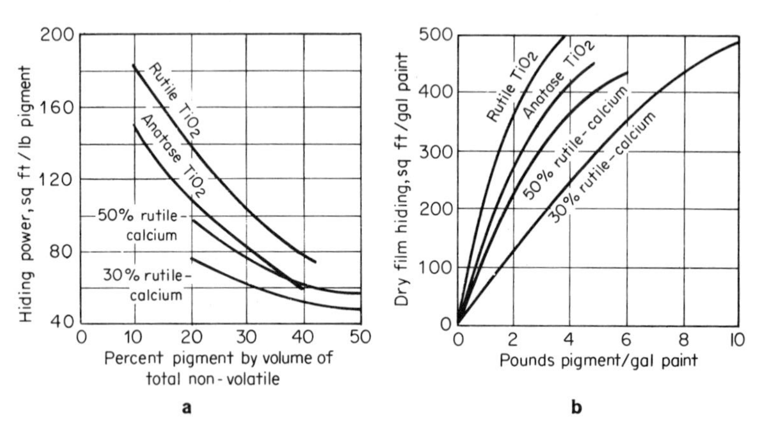

FIG. 1-1a Hiding power of TiO_2 pigments at different pigment-volume concentrations. (*Titanium Pigment Corp., subsidiary of National Lead Co., New York, New York.*)

FIG. 1-1b Dry film hiding of paints pigmented with TiO_2 pigments. (*Titanium Pigment Corp., subsidiary of National Lead Co., New York, New York.*)

TABLE 1-3 Relative Hiding Power and Tint Strength of White Pigments

	Refractive index	Hiding power, sq ft/lb	Tint strength
Rutile titanium dioxide	2.76	147	1850
Anatase titanium dioxide	2.55	115	1250
50% rutile TiO_2–calcium base	....	82	975
Zinc sulfide	2.37	58	640
Lithopone	1.84	27	280
Antimony oxide	2.19	22	300
Dibasic lead phosphite	2.25	20	250
Zinc oxide	2.01	20	210
Basic carbonate white lead	2.01	18	160
Basic sulfate white lead	1.93	14	120
Basic silicate white lead	1.83	12	80

Extracted from "The Handbook," Titanium Pigment Corp., subsidiary of National Lead Co., New York City, N.Y.

These pigments are:

Basic silicate white lead: modern, widely used type of white lead pigment

Basic carbonate white lead: still widely used, but exhibits severe chalking

Basic sulfate white lead: inferior to basic carbonate white lead; tends to cause increased film weathering

Zinc oxide (acicular form best): exhibits neutralizing properties of white lead, but does not cause film yellowing with age

Leaded zinc oxide (basic sulfate white lead and zinc oxide): a good low-cost pigment

The white lead pigments yellow and darken with age, because they react with hydrogen sulfide and other sulfur-containing gases in the atmosphere to form black lead sulfide.

A modern reactive pigment now widely employed is a combination of basic silicate white lead and zinc oxide. This combination both neutralizes film acids and increases film toughness and hardness.

3. Extenders: These pigments are added for special properties which they impart to the paint film, as well as to reduce pigment cost:

Barium sulfate (barytes, blanc fixe): used to extend the effectiveness of titanium dioxide pigments and in sanding primers for metal finishing. Exhibits good resistance to weathering

$CaCO_3$ (calcite, whiting, chalk): widely used extender because of its desirable properties—aids in reducing mold, enhances color brightness, and is low in cost

Silicates (clay, talcs, mica, diatomaceous earth): these are used to reduce gloss (clays and diatomaceous earth) and to reduce film cracking (talcs and mica)

Special-purpose extenders would include magnesium oxide, magnesium carbonate, barium carbonate, and lime. Mica is preferred over talc because talc exhibits poor mold resistance and chalks badly. Calcium sulfate (anhydrite) is not widely used because of water sensitivity, chalking, and poor mold resistance. Magnesium oxide and calcium oxide extenders should not be used in oleoresinous paints because of their reactivity with the drying oil.

4. Inhibitive pigments: These are employed in ferrous metal primers to reduce corrosion by reacting with the soluble initial iron corrosion product to form insoluble compounds:

Red lead: good weathering if topcoating must be delayed

Zinc chromate: best if topcoated soon after application

Dibasic lead phosphite: also useful as a topcoat pigment for marine and salt water exposure

Lead suboxide: a neutralizing type of inhibitive pigment for synthetic resin binders

Iron oxide often is used as an extender in metal primers.

5. Colored pigments: This class of pigments encompasses a wide variety of both organic and inorganic compounds. Pigments are not to be confused with dyes; pigments are insoluble while dyes are soluble in the paint vehicle.

The stability of pigment and dye colors is a very complex matter and has been the subject of intensive investigation by paint companies for some time. The nature of the contaminants in industrial atmospheres has a strong effect on pigments and dyes. For example, white lead paints in city atmospheres containing sulfur dioxide darken due to the formation of lead sulfide.

Recently it has been found that smog constituents have had unexpected effects on paint colors. Active gaseous chemicals such as chlorine, ozone, carbon monoxide, and sulfur dioxide may cause color changes. For example, blues tend to fade or may turn to green in oxidizing atmospheres, as in the presence of chlorine.

Great improvements have been made in the stability of pigments used in paints and coatings for exterior exposure. The phthalocyanine blues, greens, and reds are recommended, as well as the chrome yellows and molybdate oranges. These last two are not alkali-resistant, however, and a nickel azo complex yellow termed "green gold" or a parachlor red may be substituted under these conditions. In highly corrosive areas, black is the proper choice. Aluminum is too reactive to be employed in contaminated industrial atmospheres—white and gray colors should be selected which are pigmented by carbon black and chemically stable titanium dioxide pigments. Flaked stainless steel pigments are finding application in coatings used in corrosive services where aluminum shows short life.

The following listing presents commonly employed pigments for paints and coatings for exterior exposure:

WHITE:	Rutile titanium dioxide, calcium base; blends of rutile and anatase titanium dioxide (controlled chalking); titanated zinc oxide.
BLACK:	Carbon black.
ALUMINUM:	Leafing and nonleafing types.
BLUES:	Iron blues (dark); phthalocyanine blues (light).
GREENS:	Chrome greens; phthalocyanine greens.
YELLOWS:	Iron-oxide yellow; chrome yellow; green gold.
ORANGE:	Molybdate orange; organic orange "RK."
RED:	Iron-oxide red; molybdate red; parachlor red.

Paint Formulation of Oil-base Paints

A brief review of paint formulation follows. For more detailed information on this complex subject the reader is referred to the many excellent publications which are available, including those of the Steel Structures Painting Council, Pittsburgh, Pennsylvania, the Painting and Decorating Contractors of America, Chicago, Illinois, and the Titanium Pigment Corporation, New York City, New York.

Oil-base paints—"trade sales" paints or "house paints"—are formulated to provide a combination of desirable properties such as stability in the can (shelf life), ease of application, good hiding power, and fume resistance, controlled chalking, and weather resistance during exposure. To obtain these properties, a number of component materials must be combined:

- Pigments for hiding power and for adding color
- Extenders to enhance properties of pigments, improve paint life, and reduce cost
- Suspending agents for the pigment to improve shelf life
- The pigment binder, or nonvolatile resinous portion of the vehicle
- A drier to decrease dry time of an oleoresinous vehicle
- The thinner, or volatile portion of the vehicle
- A bodying or thickening agent, if required, to increase the thixotropic properties of the paint
- Antiskinning agents, mildewcides, and deodorants

Various combinations of pigments and extenders are employed, depending upon the nature of vehicle, the service condition intended, and quality. There is no "standard" or recommended combination of pigments, even for a given quality of paint. Three representative pigmentations for an exterior house paint or "trade sales" paint are presented below (as percent by weight of total pigment):

example A

10 to 20% TiO_2 (rutile and anatase combined)
20 to 40% White Lead (basic silicate preferred)
20 to 30% Zinc Oxide (acicular)
15 to 45% Extenders (whiting and mica)

example B

15% TiO_2 (rutile and anatase combined)
28% Zinc Oxide (acicular)
55% Calcium Carbonate (whiting)

example C

14% TiO_2 (anatase)
45% Leaded Zinc Oxide (35%)
32% Magnesium Silicate
9% Mica

"Fumeproof" white paints contain no white lead pigments. Increasing the TiO_2 content and decreasing the extender (whiting) of Example B would increase the service life of a paint exposed in hot humid southern climates.

The amount of pigment incorporated in a paint greatly affects the appearance of the dried paint film. The pigment loading is expressed as percent by volume of the nonvolatile portion of the vehicle—the pigment volume concentration, or PVC. Paints with higher pigment concentrations will exhibit less gloss:

High gloss enamels:	20 to 27% PVC
Gloss paints	30 to 37% PVC
Semigloss paint:	40 to 47% PVC
Flat paints:	50 to 65% PVC

As the pigment loading increases, the percent volatile in the vehicle also increases.

Formulation of water-base paints is entirely different and unrelated to the formulation and pigmentation of oil-base paints. The reader is referred to the publications cited above for information on the composition of these paints.

Chapter **2**

Color Codes, Color Standards, and the Industrial Use of Color

Introduction

Purpose: The purpose of color coding as presented in this chapter is twofold: (1) To establish the location and nature of hazardous conditions and safety devices through the application of colors with recognized safety significance; and (2) to establish by means of colors a method for identifying the contents of piping, tanks, and other containers. In addition, a short discussion is included on the industrial importance of color.

Need for color codes: The advantages of adhering to color codes and standards are manifested in minimizing errors and reducing accidents, as well as in improving operating efficiency, safety, and appearance.

Code systems used: The color codes and standards presented herein follow for the most part accepted codes of the American National Standards Institute, The National Safety Council, The American Society of Mechanical Engineers, and The Compressed Gas Association.

Esthetic importance of color: The considerable effect of color on the morale and efficiency of employees is reviewed, with recommendations for color systems to provide the desired effects.

Safety Color Code

Hazard indication: This section covers the use of colors to call attention to physical hazards, the location of safety equipment, and the identification of fire and other protective equipment. It is not the intent of this code to be in conflict with any generally accepted standards or regulations with respect to the use of colors or shapes of markers in sea or air navigation, or in railroad or highway transportation.

Code colors: This Safety Color Code is based on American National Standards Institute Standard Z53.1.-1967, to which organization credit is hereby acknowledged.

RED—*Fire protection*	Shall be used for hydrants and associated piping, sprinkler piping, and firefighting equipment and facilities (excluding fire trucks which usually are white)
GREEN—*Safety*	Shall be used for first aid and personal protective equipment and facilities
WHITE; OR WHITE AND BLACK STRIPES—*Traffic*	Shall be used for traffic striping and for housekeeping purposes, such as in food and drink areas
YELLOW—*Caution*	Shall be used to designate locations of physical hazards which might cause hazardous or painful contacts such as stumbling, falling, tripping, snagging, or catching
YELLOW AND BLACK STRIPES—*Danger*	Shall be used to attract special attention to locations of the physical hazards discussed above
ORANGE—*Alert*	Shall be used to designate dangerous parts of machines or energized equipment which may cut, crush, shock, or otherwise cause injury

BLUE—*Caution (equipment repair)*.....	Shall be used on barriers or flags placed near starting switches or valves, as a warning that the equipment is shut down for service or repair
PURPLE—*Radiation*..................	Shall be used to designate radioactive hazards

Examples of applications of the Safety Color Code are presented in Table 2-1.

TABLE 2-1 Safety Color Code Examples

RED—*Fire protection:*

- Fire alarm boxes
- Fire hydrants and fire hose cabinets
- Fire buckets or pails
- Fire extinguishers and areas on walls or supports on which they are mounted
- Fire exit signs
- Fire doors
- Water lines used primarily for fire-fighting purposes
- Danger signs
- Stop buttons for *emergency* stopping of machinery

GREEN—*Safety:*

- Safety bulletin boards
- Gas mask boxes
- First aid kits and first aid cupboards
- Stretchers
- Safety deluge showers (white and green stripes)

WHITE, OR BLACK AND WHITE—*Traffic and housekeeping:*

- Location and width of aisleways
- Stairway risers and border limit lines
- Direction signs
- Location of refuse cans
- Food and drink areas; drinking fountains

YELLOW—*Caution:*

- Corner markers for piles of stored materials
- Caution traffic signs in shop and warehouse aisles
- Coverings or guards for guy wires
- Crane hooks and pulley blocks
- Suspended or projecting fixtures which extend into normal operating areas, such as traveling conveyors, low beams, and pipes
- Handrails, guardrails, or top and bottom treads of stairways

BLACK AND YELLOW STRIPES—*Danger:*

- Dangerous curbs
- Bottom risers and top landings of industrial and public stairways
- Exposed and unguarded edges of platforms, pits, and walls
- Low pulley blocks and hooks of cranes
- Car bumpers of forklift trucks and cranes
- Pillars, posts, and columns, in areas hazardous to personnel (painted up to 5 ft off the ground)
- Dead ends of passageways and roads

TABLE 2-1 Safety Color Code Examples (Continued)

ORANGE—*Alert:*

Inside of removable guards for pulleys, shafts, chains, etc.
Inside of enclosed doors for electrical equipment, etc.
Safety start buttons

BLUE—*Caution (equipment repair):*

Barriers or flags placed conspicuously on controls of equipment shut down for service or repair, such as on:
- Valves
- Electrical controls and switches
- Scaffolding, ladders
- Elevators
- Tanks, vats, ovens, driers, boilers, etc.

PURPLE—*Radiation:*

Rooms and areas where radioactive materials are stored or handled, or which have been contaminated with radioactive materials, are to be marked with purple lines or bands.
Radiation signs are to be prominently displayed.
Disposal cans for radioactive materials and radioactively contaminated items.

Piping Identification Color Codes

Purpose: This Piping Identification Color Code is recommended for all piping systems with the exception of cross-country pipelines, buried piping, and electrical conduits.

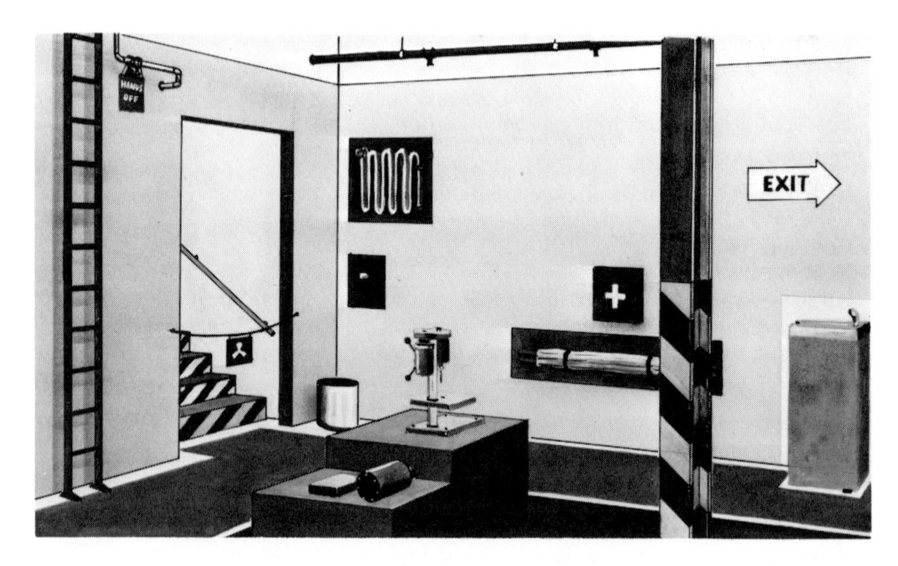

FIG. 2-1 Example of color coding for safety.

The code colors recommended below represent an extension of the standard colors sponsored by the American Society of Mechanical Engineers and the National Safety Council. The colors, legends, etc., are based on—and extracted from—the "American Standard Scheme for the Identification of Piping Systems" (American National Standard A13.1-1956) with the permission of the publisher, the American Society of Mechanical Engineers, 345 East 47th Street, New York, New York, 10017.

Because there is a limited number of colors which are readily distinguishable under all conditions of lighting and aging, materials having some property in common are grouped together and represented by one base or ground color. Within each color group, specific materials are indicated by a secondary color which may be applied as a stripe or used for letters in a stencil.

For reasons presented below, aluminum has been reserved as a nondesignating color except in a few instances where it is applied as a secondary color stripe upon a base color.

In all cases where identification is essential to safe operation, the addition of a legend is mandatory and the use of color coding becomes optional. However, when color coding is desired in addition to legends, the system presented herein is recommended.

Aluminum color: Since it is often undesirable to paint the entire length of the many pipes in a plant with the base or ground color, aluminum has been reserved as a nondesignating color. In this case, the base color is indicated by wide bands.

The base colors, listed below, have been selected as being distinguishable under poor and varying light conditions:

RED..............	Fire-protection materials
WHITE............	Steam (all pressures)
YELLOW...........	Chemicals and dangerous materials
GRAY.............	Crude oil, lube oil
ORANGE...........	Volatile petroleum products (mineral spirits and lighter)
OXIDE RED.......	Nonvolatile petroleum products (kerosine and heavier, including waxy distillates and diesel gas oil)
BLACK............	Residual oils, still bottoms, slop oils, and asphalts
BLUE..............	Water (all purities and temperatures)
GREEN............	Air and its components, and Freon

Secondary colors: The narrow stripe presenting the secondary color which identifies the specific material, may be applied by painting or preferably by use of adhesive plastic tapes of the correct color. These tapes, manufactured for this purpose, are available from many vendors

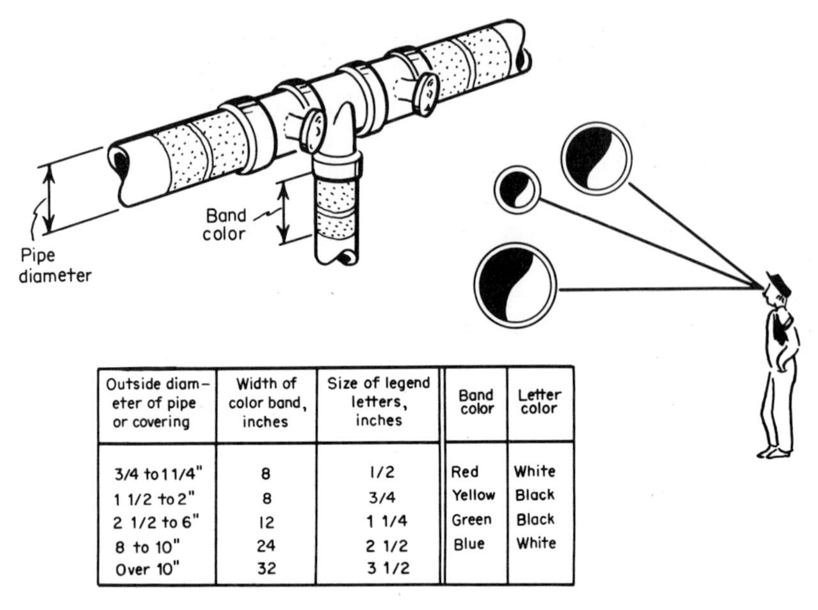

Outside diameter of pipe or covering	Width of color band, inches	Size of legend letters, inches	Band color	Letter color
3/4 to 1 1/4"	8	1/2	Red	White
1 1/2 to 2"	8	3/4	Yellow	Black
2 1/2 to 6"	12	1 1/4	Green	Black
8 to 10"	24	2 1/2	Blue	White
Over 10"	32	3 1/2		

FIG. 2-2 Specifications of color bands for identifying piping contents: size and location of bands, stripes, and letters.

with colors matching the American National Standards Institute colors. In addition, the contents of the pipe and/or direction of flow may be further identified by arrows and legend, and *must* be identified by legend if a hazard is involved.

Color bands: The location and size of bands and stripes, as recommended by American National Standard A13.1-1956, are presented in Fig. 2-2. Bands, and stripes when used, shall be applied to the pipe:

On each side of and adjacent to valves and tees and other fittings of importance

Where the pipe enters and emerges from walls and where it emerges from road and walkway overpasses

At uniform intervals along long sections of the pipe

Adjacent to tanks, vessels, and pumps

The band width and size of letters in legends will depend to some extent upon the pipe diameter. Either white or black letters are selected to provide maximum contrast to the band color. Stripes usually are 2 in. wide regardless of band width and are spaced 1 in. apart when two stripes are employed.

Examples of application of the complete Piping Identification Color Code, including both band and stripe colors, are presented in Table 2-2.

TABLE 2-2 Piping Identification Color Code

	Base color (band)	Identifying color (stripe)
Steam:		
400 psi and over	White	1—yellow
150 to 400 psi	White	1—red
Below 150 psi	White	1—green
Hot water (150°F and over):		
Distilled, demineralized	Blue	2—aluminum
Raw	Blue	2—black
Condensate	Blue	2—white
Treated (any process)	Blue	2—oxide red
Sea, salt, or brine	Blue	2—yellow
Water (below 150°F):		
Distilled and demineralized	Blue	1—aluminum
Raw	Blue	1—black
Chilled	Blue	1—gray
Condensate	Blue	1—white
Treated (any process)	Blue	1—oxide red
Sea, salt, or brine	Blue	1—yellow
Fire protection:		
Water, foam, or other fire-extinguishing material	Red	
Petroleum products:		
Gasoline (regular; leaded)	Orange	1—black
Gasoline (premium; leaded)	Orange	1—blue
Gasoline (white)	Orange	1—white
Gasoline (aviation 100/130)	Orange	1—red
Gasoline (aviation 115/145)	Orange	1—purple
LP Gas	Orange	1—oxide red
LP Gas (refrigerated)	Orange	1—yellow
Naphthas (all)	Orange	1—green
Mineral spirits	Orange	1—green
Kerosine	Oxide red	1—green
Jet fuels	Oxide red	1—yellow
Diesel oil (white)	Oxide red	1—white
Diesel oil (black)	Oxide red	1—black
Desulfurizer reactor—effluent stream	Oxide red	1—aluminum
Fuel oils	Black	1—yellow
Asphalt	Black	1—white
Slop and waste oils	Black	1—orange
Lube and crude oils:		
Lube oils	Gray	1—green
Sour crude	Gray	1—orange
Sweet crude	Gray	1—red

TABLE 2-2 Piping Identification Color Code (Continued)

	Base color (band)	Identifying color (stripe)
Chemicals:		
Acetylene	Yellow	1—purple
Acids	Yellow	1—red
Ammonia	Yellow	1—blue
Caustics	Yellow	1—black
Chlorine	Yellow	1—oxide red
Inhibitors	Yellow	1—aluminum
Sodium phosphate	Yellow	1—orange
Sodium sulfide	Yellow	1—green
Sodium sulfite	Yellow	1—gray
Sulfur dioxide	Yellow	1—white
Air and other gases:		
Hydrogen	Orange	1—aluminum
Gas (fuel and sour)	Orange	1—gray
Gas (sweet)	Orange	1—aluminum
Air (industrial)	Green	1—gray
Air (instrument)	Green	1—black
Nitrogen	Green	1—orange
Oxygen	Green	1—white
Carbon dioxide	Green	1—red
Freon	Green	1—yellow
Acetylene, ammonia, chlorine, sulfur dioxide: see chemicals		

FIG. 2-3 Example of color coding for contents of piping.

Proposed Standard Colors for Exteriors of Storage and Engine Fuel Tanks

Purpose: This section presents color standards for painting the entire exterior surface of storage and engine fuel tanks, and in addition includes a proposed contents identification color code when such identification is desired. This identification color code is consistent with the base colors for piping presented in the immediately preceding discussion.

Color requirements: All crude oil, petroleum products, and water storage tanks (excluding asphalt tanks) shall be painted in the following colors:

Roof and shell, including pipe fittings:	Aluminum
3-ft splash band around shell base, and fittings:	Black
Stairs, platforms, and ladders:	Black
Stair risers:	Yellow

Asphalt storage tanks are painted black on all surfaces including pipe fittings. Spheroids are painted as are tanks: the entire sphere is aluminum, and the supporting base is black.

Tank number and contents: At 90° intervals around the perimeter of the tank starting adjacent to a manway, the tank number and contents shall be painted in black letters on an aluminum background. Also, the date of painting should be stenciled in small letters near the manway closest to the ladder.

Identification colors: In addition to the lettering, a base-color identification band may be desired which identifies the class or group of product contained in the tank. This band is 12 in. wide and located immediately above the black splash band. The identifying color of this band conforms to the Piping Identification Color Code, as follows:

ORANGE..........	Volatile petroleum products (mineral spirits and lighter)
OXIDE RED.......	Nonvolatile petroleum products (kerosine and heavier)
BLACK............	Residual oils, asphalt
GRAY.............	Crude oil, lube oils

When it is desired to identify further the specific material within the above group of products, a secondary color may be applied as dashes on the band. Each dash shall be the full width of the band (12 in.). They shall be 2 ft long and spaced every 6 ft; that is, 6 ft of base color shall show between the dashes.

Engine fuel tank colors: In contrast to storage tanks discussed above, engine fuel tanks shall be painted over all surfaces with the identifying base color. The specific fuel contained in the tank shall be specified

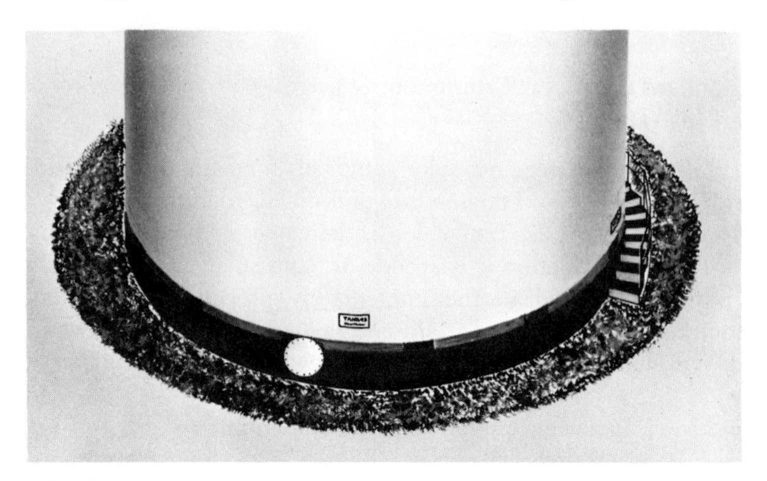

FIG. 2-4 Example of colors applied to identify contents of storage tanks.

by legend. The color of the letters can correspond to the secondary colors established in this section if so desired.

Listed below are examples of common tank fuels with their identification color code:

Fuel Tank Color Code

Fuel	Tank color	Lettering color (optional)
Gasoline (regular, leaded)	Orange	Black
Gasoline (premium, leaded)	Orange	Blue
Gasoline (white, unleaded)	Orange	White
Diesel fuel (white)	Oxide red	White
Diesel fuel (black)	Oxide red	Black
Kerosine	Oxide red	Green
Fuel oil	Black	Yellow

Examples of application of Tank Identification Color Code: The complete identification color code for storage tanks presented in Table 2-3 conforms to that for piping.

Safety Systems and Identifying Colors for Gas Cylinders

Relying upon colors alone as a means of identifying the contents of compressed gas cylinders has not been recommended. According to the Compressed Gas Association, there are over 90 types of compressed

TABLE 2-3 Storage Tank Identification Color Code

Petroleum product	Base color (band)	Secondary color (dash)
Gasoline (regular, leaded)	Orange	Black
Gasoline (premium, leaded)	Orange	Blue
Gasoline (white)	Orange	White
Gasoline (aviation, 100/130)	Orange	Red
Gasoline (aviation, 115/145)	Orange	Purple
Naphthas (all)	Orange	Green
LP Gas	Orange	Oxide red
Kerosine	Oxide red	Green
Jet fuels (all)	Oxide red	Yellow
Diesel oil (white)	Oxide red	White
Diesel oil (black)	Oxide red	Black
Desulfurizer reactor effluent	Oxide red	Aluminum
Fuel oils	Black	Yellow
Slop and waste oils	Black	Orange
Asphalt (cutbacks)	Black	White
Lubricating oils	Gray	Green
Sour crude oil	Gray	Orange
Sweet or stabilized crude oil	Gray	Red

gases distributed in gas cylinders. This is too great a number to be identified by combinations of the six primary colors, plus black and white.

Because of the serious consequences usually associated with the accidental mixing of compressed gases, three identifying or control methods are recommended by the Compressed Gas Association:

1. Primary identification through labels (refer to American National Standard Z48.1-1954; CGA Pamphlet C-4 "Method of Marking Portable Gas Containers to Identify the Material Contained")
2. Secondary safety system against mixing gases through use of special valve outlet threads (refer to American National Standard B57.1-1965; CGA Pamphlet V-1 "Compressed Gas Cylinder Valve Outlet and Inlet Connections")
3. Use of identifying colors on the cylinders of gases, considered of minor importance.

Labels: Compressed gas cylinders shall be marked by stencilling, stamping, or labeling with the commonly accepted name of the contents. This shall be located on the shoulder of the cylinder, using letter heights at least $1/25$ of the diameter of the cylinder or $1/8$-in. minimum. For international trade, the chemical symbol also is shown.

Valve outlet thread standards: The threaded outlets of the cylinder valves are separated into four basic divisions: female threads, male threads, right-hand (RH) threads, and left-hand (LH) threads. Further variations are derived by different thread diameters and pitches.

As far as possible, the assignment of valve threads to gases was made so as to prevent the interchange of connections which would produce a hazardous mixture of gases. For example, right-hand threads are used for nonfuel gases, and left-hand threads for fuel gases.

In Table 2-4 are presented the valve outlet threads for a number of commonly used gases.

TABLE 2-4 Compressed Gas Cylinder Valve Outlet Threads

Compressed gas	Symbol	Valve outlet threads				Notes
		Diameter, in.	Threads/in.*	Turn	Type	
Acetylene	C_2H_2	0.885	14 NGO	LH	Female	
Air (industrial)	AIR	0.965	14 NGO	LH	Female	
Ammonia	NH_3	⅜	18 NGT	RH	Female	
Argon	Ar	0.965	14 NGO	RH	Female	
Butane	LPGas	0.885	14 NGO	LH	Female	
Carbon dioxide	CO_2	0.825	14 NGO	RH	Male	Flat nipple
Chlorine	Cl_2	1.030	14 NGO	RH	Male	Without groove
Freon 12	CCl_2F_2	1.030	14 NGO	RH	Male	With groove
Freon 22	$CHClF_2$	1.030	14 NGO	RH	Male	With groove
Helium	He	0.965	14 NGO	RH	Female	
Hydrogen	H	0.825	14 NGO	LH	Male	Round nipple
Hydrogen sulfide	H_2S	0.825	14 NGO	LH	Male	Flat nipple
Nitrogen	N_2	0.965	14 NGO	RH	Female	
Oxygen (industrial)	O_2	0.903	14 NGO	RH	Male	
Oxygen (medical)	O_2	0.745	14 NGO	RH	Male	
Sulfur dioxide	SO_2	1.030	14 NGO	RH	Male	With groove

* NGO: National Gas Outlet thread. NGT: National Gas Taper thread.
Reference: American National Standard B57.1-1965, sponsored by the Compressed Gas Association: CGA Pamphlet V-1 (1965).

Color systems: Although relying on colors alone is not satisfactory for the identification of compressed gases, colors are useful for purposes of cylinder segregation, ownership identification, and to serve as a warning to the user when a change in cylinder color is noticed. The cylinder colors employed by a number of the major suppliers of industrial compressed gases conform for the most part to those presented in Table 2-5.

The medical profession has relied to a greater extent upon the use of colors for the identification of compressed gases. The International Organization for Standardization (ISO) has proposed the use of the

TABLE 2-5 Compressed Gas Cylinder Colors in Common Use

Compressed gas	Symbol	Body color	Shoulder color
Acetylene	C_2H_2	Black	
Air	AIR	Blue	
Ammonia	NH_3	Black	Gray
Argon	Ar	Green	Orange
Carbon dioxide	CO_2	Silver	
Chlorine	Cl_2	Red	
Freon 12	CCl_2F_2	White	
Freon 22	$CHClF_2$	White	
Helium	He	Orange	
Hydrogen	H_2	Red	Black
Hydrogen sulfide	H_2S	Blue	Gray
Liquefied petroleum gas	LPGas	Aluminum	
Nitrogen	N_2	Black	Orange
Oxygen (industrial)	O_2	Green	
Oxygen (medical)	O_2	Green	White
Sulfur dioxide	SO_2	Yellow	

identifying colors for medical compressed gases as presented in Table 2-6. These colors conform to considerable extent to those employed in the United States, which also are presented in Table 2-6 for comparison.

TABLE 2-6 Color Standards (Proposed) for Anesthetic Gas Cylinders
(International Organization for Standardization [ISO])

	ISO Standard R-32		U.S.A. practice	
Medical gas	Body color	Shoulder color	Body color	Shoulder color
Air	Gray	Black/white		
Carbon dioxide	Gray	Gray	Gray	Gray
Cyclopropane	Orange	Orange	Orange	Orange
Ethylene	Violet	Violet	Red	Red
Helium	Brown	Brown	Brown	Brown
Nitrogen	Gray	Black		
Nitrous oxide	Blue	Blue	Blue	Blue
Oxygen	Black	White	Green	Green

The Industrial Importance of Color

In addition to the use of color for purposes of safety and of identification of the contents of piping and containers, color has become an increasingly important factor in industrial painting because of its strong psychological effect on employees and customers. It has been amply demonstrated that the proper choice of color will:

Reduce eye and body fatigue

Reduce accident frequency and type

Improve morale, housekeeping, and care of equipment

all of which will result in increased quality and quantity of production. A brief review of color theory will aid in the understanding of the application of color to obtain these desired effects.

Three independent factors determine a color as we see it. These are:

1. *Hue*—the *chromatic* quality which determines the basic color type which we name, such as red, blue, etc.

2. *Value*—the *lightness* of the color in relation to white or black, which we call the "shade" of the color, such as light blue, dark green, etc. The value of a color is adjusted by adding white or black to the base color.

3. *Intensity*—the *color strength* of the color, sometimes called the purity of the color, such as bright blue, muddy yellow, etc. The intensity of a color may be downgraded by adding a gray of the same value, which neither lightens nor darkens the color.

It has been determined that most people can recognize twelve colors when viewed alone. These are the six "basic" colors: red, orange, yellow, green, blue, and violet, and the six intermediate colors orange-yellow, yellow-green, etc. These colors often are arranged in a "color circle" as shown in Fig. 2-5 so that the "opposite" or complementary colors are diametrically opposed on the circle.

The complementary color to any chosen color results when the chosen color is subtracted from white. This also may be illustrated by staring fixedly at a color for a minute or two, then shifting the eye to a gray or white surface. The complementary color will appear before the vision for a short time.

Pleasing color combinations will result if chosen as follows:

1. Use of complementary colors: painting minor areas with shades of color complementary to the color in principal use

2. Use of adjacent colors on the color circle: painting minor areas with shades of the two colors adjacent to the major color will increase the richness of the color scheme

3. Use of color triads: a three-color scheme through selection of colors each 120° apart on the color circle.

Colors have a strong psychological effect on the human being. Bright and strong colors act as physical stimulants. They increase muscular tension, blood pressure, etc., leading to increased bodily activity. Dull cool colors, on the other hand, are relaxing, and so contribute to mental

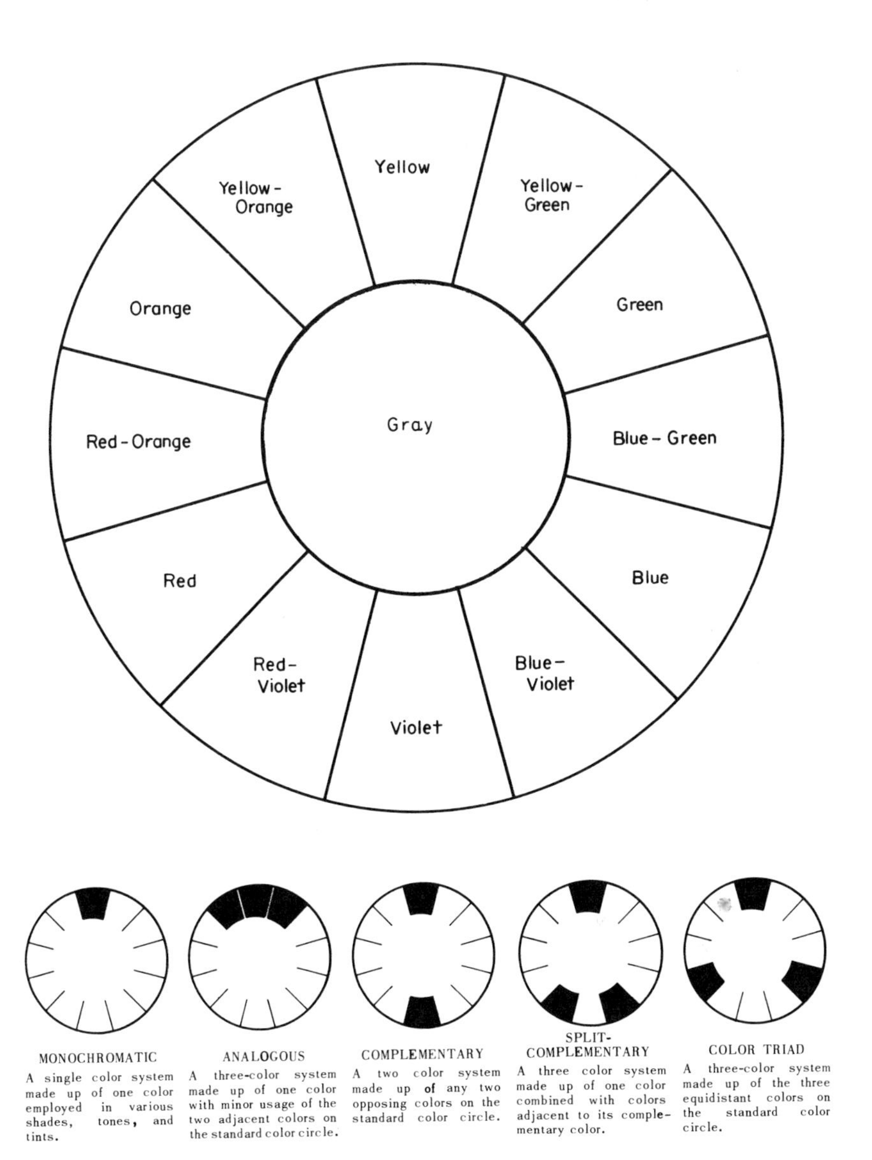

FIG. 2-5 Standard color circle and color systems.

activity. Specific psychological effects are:

YELLOW:	Happy, cheerful
ORANGE:	Warm, cheerful
RED:	Passionate, exciting
VIOLET:	Wistful, somewhat neutral
BLUE:	Cool, quiet
GREEN:	Calm, somewhat cooling
GRAY:	Neutral, no effect
BLACK:	Depressing

Black is considered a warm color, but should be used only on small areas; otherwise it becomes depressing. Black combines well with reds and oranges. White is a cool, spirit-lifting color, and may be used on large areas for psychological reasons as well as to increase the light intensity. It combines well with greens, blues, and violet.

The choice of color combinations greatly affects visibility. The legibility of colored signs at considerable distance was determined to be as follows, in order of decreasing legibility:

1. Black on yellow
2. Green on white
3. Red on white
4. Blue on white
5. White on blue
6. Black on white
7. Yellow on black
8. White on red
9. White on green
10. White on black
11. Red on yellow
12. Green on red
13. Red on green

The above results will hold some surprises for the reader. For example, the commonly employed black letters on a white background are listed sixth in order of visibility.

In the industrial application of color, particular attention also must be paid to the gloss or shininess of the paint applied, and to the effect of the color on the light intensity used in illuminating working areas. Eye fatigue, which has a direct bearing on body fatigue, will result from glare, excessive color contrasts in the field of vision, and constant pupillary adjustments from light to dark.

Consequently, color schemes for factory and office should be kept simple, pastel in shade, and flat or semigloss in sheen. To avoid unnecessary contrasts girders, piping, supports, etc., should be painted in the same color as the walls. White is best for reflectivity and is recommended for use on ceilings, and may be extended down the walls to the illumination level.

The following is a direct quotation from a paper presented by the late Mr. Faber Birren, industrial color consultant, before the National Research Council of the National Academy of Sciences in Washington, D.C. in 1959 (Publication 653: "Field Applied Paints and Coatings"):

> Bright general illumination and color tend to stimulate an *outward* attention toward the environment. Because of excitation, it prods the body and puts it on the alert, visually and physically. Interest is drawn to surroundings. The bright environment, with pale color on walls (yellow, light green) is therefore appropriate for many industrial operations such as machine shops, large assembly, metalwork, and woodwork.
>
> Soft colors on walls tend to minimize the environment as a source of interest. Here there will be more of an *inward* attitude and attention. The softness of the walls will not demand visual attention. Hence, the worker will be better able to concentrate on difficult visual and mental tasks—bookkeeping, fine parts assembly, fine tool making, inspection. General illumination should be kept within reasonable limits, and supplementary light may be concentrated directly over the task where necessary. Ideal wall colors will be soft tones of gray, green, or coral.
>
> All things considered, the following notes offer general recommendations for use of color in industry. By and large, ceilings and overhead should be white, with a reflectance of 75 percent or better. Colored overhead is undesirable because it tends to quench light reflection, and because it may pull visual and emotional interest up and away from the vital working areas and machinery of the plant. Upper walls in areas devoted to *manual* tasks (such as machine shops) may have a reflectance to 60 percent or slightly more. Good colors are yellow, ivory, or pale green. In areas devoted to difficult *visual* or *mental* tasks (such as fine parts assembly or inspection) the wall brightness should be between 50 and 60 percent. Dadoes and trim, for practical maintenance, should have a reflectance between 25 and 35 percent. Dado and trim colors are preferable in deeper shades of the wall tint to avoid too much color contrast. Machinery and equipment should likewise reflect between 25 and 35 percent.
>
> Bright colors should not be applied to relatively meaningless objects such as lockers, shelving, bins, racks. Gray is suggested.
>
> What colors should be used in industry, and how many? First of all, all colors except white or pale yellow should be slightly grayed to keep them from being too vivid and distracting and to stand up better in maintenance. Grayed colors do not show soiling as much as "clean" colors. Green is one of the best of all hues, fresh in appearance and slightly passive in quality. Blue-green is also excellent, being the

direct complement of the average human complexion. Cool colors, such as green and blue-green, serve the added purpose of offering psychological compensation for high temperatures so commonly found in industrial plants.

While yellow-green reflects light in the high visibility region of the spectrum, it tends to look "bilious" and its reflections, if any, on human flesh are unpleasant. Yellow is good for large, vaulty spaces. Tones of buff and coral are quite suitable for facilities devoted to women, or for a desired "warmth" to compensate for chilly room temperature. Gray, being neutral in emotional quality, is useful in a restricted way for areas in which the tasks require severe visual or mental concentration and in which color may compete for attention. For the most part, blue is not a good functional color. The light energy it reflects also may cause the eye to be nearsighted and hence to distort the focus of the eye, even though slightly.

In writing color specifications for industrial plants, the following suggestions may be helpful. Room *purpose* is far more important than orientation. That is, while an interior decorator may suggest cool colors for south exposures and warm colors for north exposures, this device holds less significance in an industrial plant than the requirements of the job to be done. Having functions rather than appearances in mind, soft green and blue-green are recommended for work areas in general, machine shops, etc. They are also suitable for offices, inspection areas, and wherever most of the employees are concentrated at average tasks.

Where there may be total lack of daylight, and where a warm rather than cool environment may seem preferable, tones of buff, peach, or coral may be applied. Corridors and stairways are best in a luminous color such as yellow. Storage areas should be plain white, both to reflect abundant light and to permit emphasis of color in other and more important areas where major work is done. In rest rooms and wash rooms, colors different from those used in production areas should be considered. Here it may be desirable to use tints of pink for women, turquoise for men, and to introduce variety and an interesting "change of pace."

Functional color offers tremendous benefits in work production. Investment in it pays many tangible and intangible dividends. It is admittedly difficult, however, to gather reliable data on improvement in work output. One carefully supervised study has been reported by the U.S. Public Buildings Administration and the U.S. Public Health Service in Washington, D.C. Over a period of two years, an analysis was made of worker efficiency among employees using business machines. The test conditions included: (a) the original room, (b) the addition of lighting fixtures only, and (c) the further addition of color. The original area, in brown and buff and with olive drab equipment, was repainted in light tones of green and gray.

As to worker efficiency, one task had an improvement of 37.4 percent. However, a conservative figure of 5.5 percent has been set as the general improvement shown in the department. In cash value, this 5.5 percent production improvement was equivalent to a saving on gross payroll of $13,229 among some 95 employees. If this figure is a credible one, right illumination and right color are worth about $139.25 annually per average employee, $13,925 with 100 employees, $139,250 with 1,000 employees. The monetary value of color is substantial and should be looked on as a sound investment.

The exterior application of color in industrial plants and refineries also is of growing interest. Many of the principles in the preceding discussion which was concerned with plant and office interiors also apply to the exterior use of color. In addition, however, the stability of the color in the exterior exposure now must be considered.

The subject of colored pigments in paints and coatings for exterior exposure is reviewed briefly in Chap. 1 under the section, "Colored Pigments," on page 36.

Chapter **3**

Safety and Health Considerations; Smog Control Legislation

Introduction

This chapter outlines the safety and health hazards that must be considered when preparing a surface for painting and when applying paint, and indicates the precautions required to reduce these hazards. Also included is a short discussion on the importance of smog control legislation as it affects paint formulation and application.

Special Hazards: In addition to the usual hazards associated with construction activities, workmen engaged in surface preparation and paint application are exposed to dangers of fire, explosion, burns, toxic fumes, dust, and insufficient fresh air. The damage that may result from the explosion of a small quantity of vaporized solvent, and the dangers to health inherent in fume and dust inhalation should be fully realized by the workmen and their supervisors. Whenever solvents are present, care must be taken to provide sufficient ventilation to prevent the accumulation of combustible concentrations of fumes and that air-supplied respirators or hoods are worn when fume concentrations become toxic.

Supervisory Precautions: The supervisor shall see that prior to starting any type of surface preparation or paint work, suitable precautions have been taken by the workmen to prevent possible injury and that a work permit has been obtained as required.

Before entering enclosed spaces such as tanks, vessels, or other areas known or suspected to contain gaseous concentrations harmful to life, the supervisor must determine their concentration with a combustible gas indicator. Other sampling devices to determine toxicity, presence of harmful dusts, etc., must be used as required. No work permit will be issued to enter a confined space with a dangerous gas concentration unless special prescribed precautions are met.

The selection of recommended protective equipment and apparel to be worn during various methods of surface preparation and paint application is presented in Table 3-8 on page 77.

FIG. 3-1 A combustible gas indicator. (*Mine Safety Appliances Co., Pittsburgh, Pennsylvania.*)

Hazards During Surface Preparation

Hazards from tools and equipment: Wire-brushing, chipping, scraping, and grinding methods used in preparing a surface for painting produce flying particles or chips which can maim or blind. Disintegrating abrasive wheels and breaking wires from rotary wire brushes are important hazards associated with power-tool cleaning.

All power tools used in surface cleaning should be inspected for good operating condition before use. Cracked grinding wheels and worn wire brushes are to be discarded and not used. All power equipment should be operated at speeds recommended by the manufacturer. Where required, all tools are to be effectively grounded. Filter-type respirators must be worn where dust hazards are present.

Blast cleaning operations (sandblasting, shotblasting, etc.) are particularly hazardous, and all required precautions must be taken to avoid accidents. The blast stream from a sandblasting nozzle is in effect a continuous-action shotgun and is potentially just as dangerous. Pneumatically operated "deadman"-type remote control systems, more than any other single factor, have helped eliminate the hazards involved. Remote control systems enable the blaster to stop and start the machine at the nozzle, and should the blaster accidentally drop the hose the operation will automatically stop.

Fire hazards can sometimes be a problem, especially when working in confined areas within chemical or petrochemical plants. Some abrasives can cause sparking when the particles strike the surface. Usually the large quantities of air expelled through the nozzle will rapidly dis-

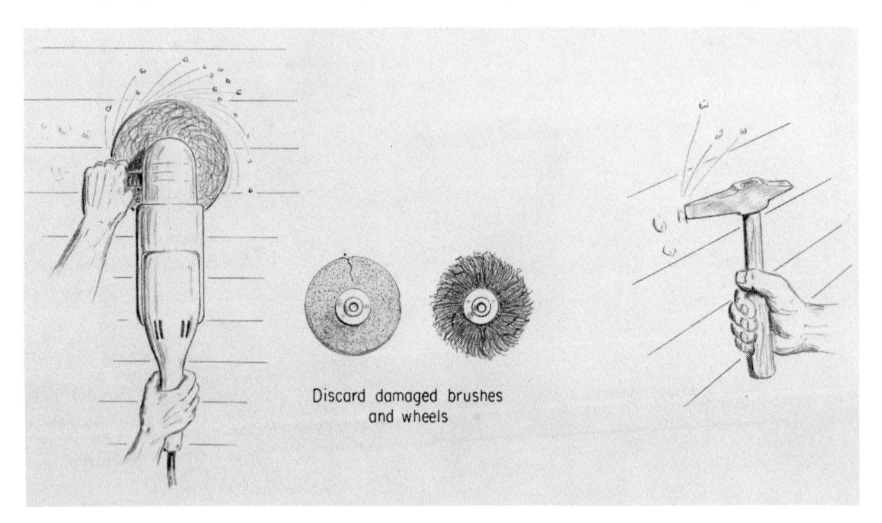

FIG. 3-2 Examples of hazards when using hand and power-cleaning tools.

FIG. 3-3a Sandblasting hazard without a deadman control.

sipate any combustible vapors in the immediate blasting area. There is a possibility, however, of ricocheting particles falling in combustible areas some distance from the air stream. This could be a potential explosion or fire hazard. When inside confined areas of any type such as pipes or tanks, they must be proven gas free before work is started.

It is equally important to have the sandblast nozzle effectively grounded, so as to discharge the relatively high static electricity charges that are developed.

Protective devices for the eyes, skin, and respiratory systems must be used in all sandblasting operations. Air-supplied hoods provide pro-

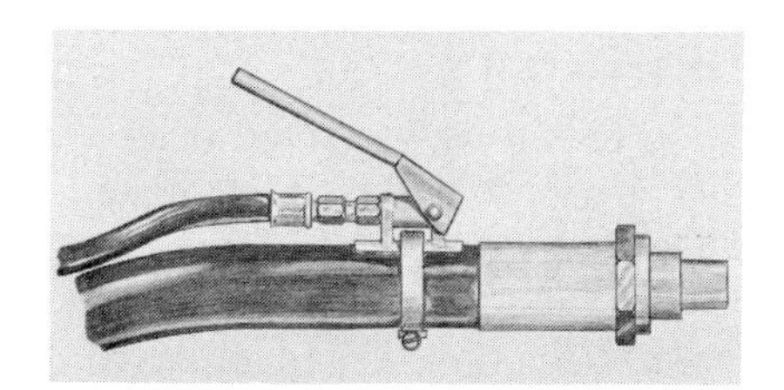

FIG. 3-3b A representative sandblast deadman control.

FIG. 3-4 Typical sandblast safety apparel. (*Mine Safety Appliances Co., Pittsburgh, Pennsylvania.*)

tection against dust inhalation. Dust inhalation over a period of years can cause serious respiratory diseases such as silicosis.

Scaffolding, Platforms, Ladders, Ropes: Care must be taken to protect workmen against accidents when using or working on scaffolding, platforms, ladders, or in any elevated position. Scaffolds and platforms shall be provided with a backrail to prevent workers from falling.

All movable scaffolding, including ropes, blocks, hooks, and ties, shall be examined every day before use. Any apparent or suspected defect or unsafe condition shall be corrected immediately. Scaffold ropes must be wrapped to prevent dangerous weakening from blast cleaning or from spotting by acid and alkali. Rope, hose, and equipment shall be located so as not to endanger workmen.

All extension and straight ladders shall be equipped with safety shoes. Ladders shall be set level at the base and given sufficient bearing. All ladders shall be inspected once a month. Ladders and scaffolds preferably are of the type recommended by the National Safety Council.

General Precautions: No surface preparation or painting work shall be performed in the immediate vicinity of moving parts such as drive shafts, belts, gears, and pulleys. When working near such equipment,

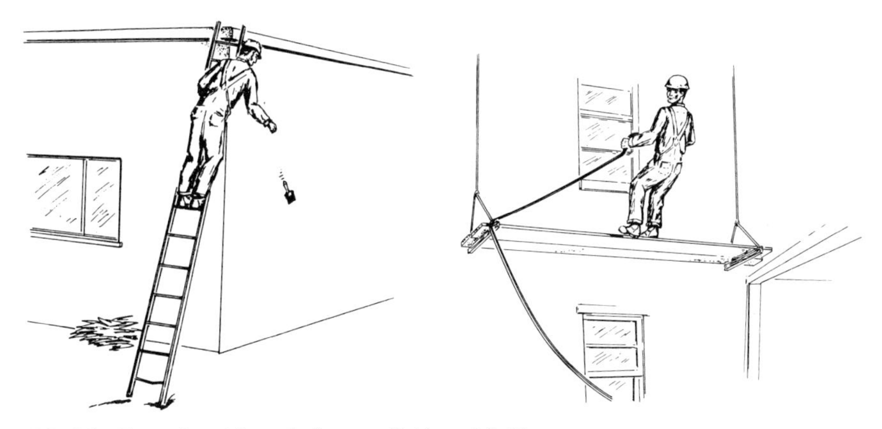

FIG. 3-5 Examples of hazards from scaffolds and ladders.

care shall be taken to prevent the accidental starting of the equipment, and blue warning flags shall be placed on controls (see Chap. 2: "Safety"). All piping coming from or leading to a vessel or tank shall be secured by blinds, double block valves with a bleeder, or a section of line removed before workmen are allowed to enter.

Safety belts with lines attached should be worn when working inside any tank or small enclosure or on top of high equipment such as tanks, structural steelwork, and towers. When working in a tank, the attached lines shall extend through a manhole or other opening, and an observer shall be posted outside to watch the operator and to be prepared to help remove him to safety when necessary. Safety belts must be tested once a month.

All electrical equipment and connections used near continuous painting operations shall be of the explosion-proof type. Sources of flame and sparks (such as matches, smokes, welding operations, etc.) must be eliminated when flammable gases may be present. When possible, commutator-type electric motors are to be shut down during paint applications in their vicinity. Sparkproof tools shall be employed, and rubber-soled shoes are recommended for the workmen.

Hazards from generated dusts formed during surface preparation: Dusts generated during surface preparation can result in serious health hazards. Toxic dusts may cause systemic poisoning, and dermatitis can result from irritating dusts deposited on skin surfaces.

The recommended maximum concentrations of several types of hazardous dust particles that may be inhaled without harm during an 8-hr period have been determined by the American Conference of Governmental Industrial Hygienists. These values, termed the "Threshold Limit Values" (TLV), are presented in Table 3-1 with their permission.

TABLE 3-1 Threshold Limit Values for Toxic Dusts

Substance	*Threshold Limit Values (millions of particles/ cu ft air)*
Silica: quartz and cristabolite	$250 \div (5 + SiO_2\ \%)$
Silica: amorphous (diatomaceous earth)	20
Silica: tremolite	5
Silicate: asbestos	5
Silicate: mica, soapstone, talc	20
Silicate: portland cement	50
Graphite (natural)	15
Inert or nuisance dusts	50*

* or 15 Mg/cu meter, whichever is the smaller.

Note: Millions of particles per cu ft × 35.3 = particles per cubic centimeter.

SOURCE: American Conference of Governmental Industrial Hygienists, "Threshold Limit Values for 1968," Cincinnati, Ohio.

The dust particle concentrations in this table were based on impinger samples counted by light-field techniques.

As a basis for judging the density of the dust concentrations presented in Table 3-1, the average concentrations of dust particles in various atmospheres are presented below:

Substance	*Approximate concentration (millions of particles/cu ft air)*
Pollen	0.05–0.2
Dust, rural air	0.2–2.0
Dust, city air	0.5–5.0
Dust, industrial district	1.0–20.0
Dust, dust storm	2,000–30,000

Proper respiratory equipment should be worn at all times when surface preparation generates dust. The following are recommended types of respirators to be worn under different dust conditions:

Up to 10X TLV: Half-face-mask dust filter
Between 10X and 100X TLV: Full-face-mask dust respirator
Over 100X TLV: Air-supplied hood or respirator

Blast cleaning operations always require the use of air-supplied hoods. These should be well made of wear-resistant materials.

Fire and Health Hazards from Paint and Solvents

The person applying paints and coatings is exposed to several health hazards related to the material being applied. Fire or explosions may

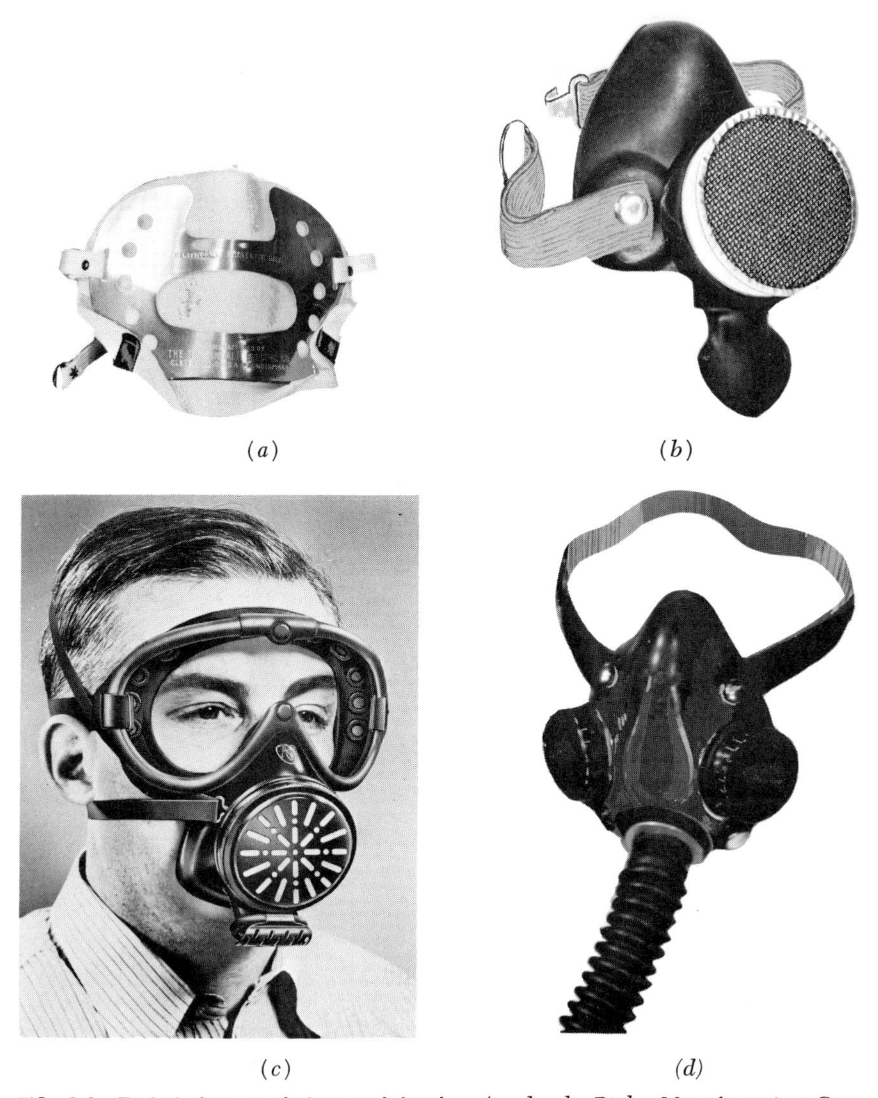

(a) (b) (c) (d)

FIG. 3-6 Typical dust respirators and hoods. *(a, b, d—Binks Manufacturing Co., Chicago, Illinois.) (c—American Optical Co., Safety Products Division, Southbridge, Massachusetts.)*

occur from careless handling of paints with flammable solvents; skin rashes and dermatitis can be caused by contact with solvents and catalysts; and various serious systemic effects can result from inhaling solvent vapors.

Fire hazards from paint solvents: Two unrelated types of fire hazards must be considered: flammability, and the likelihood of an explosion. The flammability of a liquid may be determined by several methods, but for the painter the "open cup" method is most significant.

The open-cup flash point of a flammable liquid is the lowest temperature at which the liquid (in a standard open container) will give off sufficient vapor to ignite momentarily when a flame is applied. Nearly all organic solvents contained in paint or used for preparatory cleaning have flash points below ambient temperatures. Generally, the more volatile the flammable solvent, the lower will be its flash point. The strong solvents such as toluene, xylene, and the ketones are very volatile and also very flammable. It follows then that great care always must be exercised to avoid a source of ignition.

Table 3-2 presents the flash points and boiling point ranges for most solvents of interest to painters.

TABLE 3-2 Flash Points and Flammable Limits of Paint Solvents

Solvent	Approximate boiling point range, °F	Typical flash point, open cup, °F	Explosive limits, % by vol.	
			LEL (lower)	UEL (upper)
Acetone	133–135	4	2.6	12.8
Alcohol, butyl	242–246	97–115	1.7	18.0
Alcohol, ethyl (95%)	171–174	64–68	3.3	19.0
Alcohol, isopropyl (91%)	176–178	63–67	2.5	12.0
Benzene (benzol)	172–176	12	1.4	8.0
Carbon tetrachloride	170–172	None	(nonflammable)	
Cyclohexane	176–182	32	1.3	8.0
Cyclohexanone	308–321	129	1.1	
Cyclohexene	180–183	22		
Ethylene dichloride	179–186	70	6.2	15.9
Ethylene glycol ethyl ether	274–278	116–120	2.6	15.7
Ethylene glycol ethyl ether acetate	302–321	134–139	1.7	
Ethylene glycol methyl ether	255–260	110–115		
Ethylene glycol methyl ether acetate	289–297	125–135		
Methyl ethyl ketone (MEK)	174–178	28	1.8	11.5
Methyl isobutyl ketone (MIBK)	239–243	70	1.2	9.0
Mineral spirits	310–390	105–110	0.7	6.0
Naphtha, VM and P	245–300	40–60	0.8	6.0
Toluene (toluol)	230–232	45	1.2	7.0
Trichlorethylene	188–190	None	(nonflammable)	
Turpentine	310–340	95–100	0.8	none
Xylene (xylol)	280–284	80	1.1	7.0

Note: Refer to Table 3-4, page 70, for trade names of the glycol ether solvents.

SOURCE: The Solvents and Chemicals Companies: Central Solvents and Chemicals Co., Chicago, Ill.

The possibility of an explosion occurring with a flammable solvent is dependent upon the concentration of the solvent vapors in the surrounding atmosphere. If the concentration is too high, no explosion will occur because of lack of oxygen. If the concentration is too low, there is too little of the solvent present to ignite and explode. Therefore, there are two limits of concentration between which explosions may occur, which are termed the "Lower Explosive Limit" (LEL) and the "Upper Explosive Limit" (UEL). Painters, of course, are concerned only with the lower explosive limit, and it is essential that operations be maintained on the safe side below the minimum value (LEL).

The explosive limits of various solvents also are presented in Table 3-2. It should be remembered that the explosive limit of a liquid is not directly related to its volatility or flash point (except that both effects involve the propagation of flame).

Precautions to Prevent Fires: Tools or equipment used in painting operations may constitute a fire hazard unless steps are taken to prevent sparks and other sources of ignition.

When paint is sprayed in confined areas, even though adequate ventilation is provided, the chances of a fire or explosion being caused by a spark in a local high solvent-vapor/air concentration is always present. It is therefore imperative that people who work with solvent cleaning or painting observe the following safety precautions:

1. Smoking or use of open flames is permitted only in designated safe locations.
2. Explosion-proof electrical equipment and lights shall be used with surface preparation and paint application equipment. Commutator-type electric motors should be turned off when painting nearby.
3. Solvents or paints should be applied only to cool surfaces.
4. The safety apparel specified in Table 3-8 shall be worn.
5. Ventilating equipment shall be used when working in confined areas (reference is to Tables 3-6 and 3-7).

Health hazards from paint and solvents: Two principal health hazards are involved in the application of paints. These are skin rashes and other forms of dermatitis due to absorption of catalysts and solvents through the skin, and the usually serious toxic effects resulting from inhalation of solvent vapors.

Skin rashes and other poisonous effects will develop after short contact with the amine catalysts used with coatings such as the epoxies. The amides and amine adducts are much less aggressive in this respect. The coal-tar solvents and the fumes from hot-applied coal-tar enamels will cause severe inflammation of the skin; furthermore, periodic contact over a number of months can lead to cancer. Strong degreasing solvents such as the ketones and aromatic hydrocarbons will remove the natural skin oils, causing painfully severe skin cracking.

Undesired or toxic materials should be removed by washing with soap and water soon after contact with the skin. *Do not* use solvents for this purpose—they will penetrate into the skin carrying the toxic materials with them.

Use of gloves or protective creams usually is all that is required to prevent skin poisoning. (Refer to Table 3-8 on page 77 in this chapter for recommended safety measures.)

There is a maximum allowable concentration of the vapor of most liquids that may be inhaled during an 8-hr period without hazard to health. This allowable concentration, the Threshold Limit Value, has been determined for the more commonly employed liquids by the American Conference of Governmental Industrial Hygienists. These Threshold Limit Values (1968) for a number of solvents of interest to painters are presented in Table 3-3.

The aliphatic solvents, such as mineral spirits and naphtha, have a relatively low order of toxicity in the range of 500 ppm that may be inhaled in an 8-hr period. The stronger solvents required for protective coatings, such as the ketones and the aromatic hydrocarbons toluene and xylene, are much more toxic and have TLVs in the order of 100 to 200 ppm.

Very irritating are turpentine and methyl isobutyl ketone (MIBK). Nearly all chlorinated solvents are very toxic, causing liver damage. Benzene and carbon tetrachloride are especially poisonous and *never* should be employed as a solvent or for cleaning purposes. Also very toxic are ethylene glycol monoethyl ether and its acetate ester, and paints containing these solvents must be applied only under proper ventilation conditions. As a convenience to the reader, the trade names of the more common glycol ether solvents are presented in Table 3-4.

The toxic effects resulting from inhalation of greater quantities of solvent vapors than recommended by the ACGIH are presented in Table 3-5. The effects presented are only for moderate overdoses of the solvents. More severe exposure usually results in permanent disability or death.

Ventilation requirements: It is obvious that ventilation is required when applying paint in confined areas to lessen the concentration of flammable vapors to acceptable limits. During paint application, solvent vapors constantly are being liberated to the surrounding atmosphere. Fresh air, therefore, just as constantly must be supplied in order to limit the concentration of these vapors.

Paint application by spraying results in a more rapid generation of higher vapor/air ratios than does brush or roller application. Spray

TABLE 3-3 Maximum Allowable Concentrations of Solvent Vapors Which May be Inhaled in 8 Hr Without Toxic Effects

Solvent	*Threshold Limit Value (TLV) for 8 hr; ppm by volume*
Acetone	1,000
Alcohols	
Ethyl	1,000
Isopropyl	400
Methyl (wood)	200
Benzene (benzol)	25
Carbon tetrachloride	10
Cyclohexane	300
Cyclohexanone	50
Cyclohexene	300
Ethylene glycol butyl ether	50
Ethylene glycol ethyl ether	200
Ethylene glycol methyl ether	25
Ethylene glycol ethyl ether acetate	100
Ethylene glycol methyl ether acetate	25
Enamel thinner	200
Ethylene dichloride	50
Methylene chloride	500
Methyl ethyl ketone (MEK)	200
Methyl isobutyl ketone (MIBK)	100
Mineral spirits	500
Naphtha, coal tar	200
Naphtha, VM and P	500
Perchloroethylene	100
Toluene (toluol)	200
Trichlorethylene	100
Turpentine	100
Xylene (xylol)	100

Note: Refer to Table 3-4, page 70, for trade names of the glycol ether solvents.

SOURCE: American Conference of Governmental Industrial Hygienists: "Threshold Limit Values for 1968," Cincinnati, Ohio.

painting requires closer attention to proper ventilation requirements, especially in confined areas. In open areas the solvent vapor is dissipated so rapidly that flammable mixtures are formed only in a very small region near the spray nozzle, and so little fire hazard is present.

It is essential that the air change rate be sufficient to maintain the solvent vapor concentration safely below the Lower Explosive Limit (LEL). A safety factor of 4, representing 25 percent LEL, is desired

TABLE 3-4 Proprietary Trade Names of Glycol Ether Solvents

Common Name	Carbide*	Dow†	Eastman‡	Enjay§	Olin¶
Ethylene glycol monoethyl ether	Cellosolve Solvent	Dowanol EE	Ektasolve EE	Ethyl Jaysolve	Poly-Solv EE
Ethylene glycol monoethyl ether acetate	Cellosolve Acetate	Dowanol EE Acetate	Ektasolve EE Acetate	Ethyl Jaysolve Acetate	Poly-Solv EA
Ethylene glycol monomethyl ether	Methyl Cellosolve	Dowanol EM	Ektasolve EM	Methyl Jaysolve	Poly-Solv EM
Ethylene glycol monomethyl ether acetate	Methyl Cellosolve Acetate	Dowanol EM Acetate	Ektasolve EM Acetate	Methyl Jaysolve Acetate	
Ethylene glycol monobutyl ether	Butyl Cellosolve	Dowanol EB	Ektasolve EB	Butyl Jaysolve	Poly-Solv EB
Diethylene glycol monoethyl ether	Carbitol Solvent	Dowanol DE	Ektasolve DE	Jaysolve DE	Poly-Solv DE
Diethylene glycol monomethyl ether	Methyl Carbitol	Dowanol DM	Ektasolve DM	Jaysolve DM	Poly-Solv DM

* Union Carbide Corporation, New York City, N.Y.
† Dow Chemical Co., Midland, Mich.
‡ Eastman Chemical Products, Inc., Kingsport, Tenn.
§ Enjay Chemical Co., Division of Humble Oil & Refining Co., New York City, N.Y.
¶ Olin Chemical Co., Division of Olin-Mathieson Chemical Corp., Houston, Tex.

TABLE 3-5 Toxic Effects From Overexposure to Paint Solvents

Solvent	*Effect of inhalation*
Acetone	Irritating to mucous membranes; choking sensation
Alcohol, ethyl	Intoxication
Alcohol, isopropyl	Intoxication; headache
Benzene (benzol)	Injury to blood-forming organs, and to heart, liver, kidneys, etc.
Carbon tetrachloride	Nausea, headache, vomiting; injury to liver (nephritus)
Ethylene glycol ethyl ether	Irritating to eyes; disagreeable odor
Ethylene glycol methyl ether	Possibly affects blood-forming organs
Ethylene dichloride	Irritating to nose; retching; unconsciousness
Methyl ethyl ketone (MEK)	Irritating to nasal passages; choking sensation
Methyl isobutyl ketone (MIBK)	Irritating to mucous membranes; choking sensation
Naphtha, VM and P	Headache; vomiting; muscular twitching
Toluene (toluol)	Same as for benzene except little damage to blood-forming organs
Trichlorethylene	Similar to carbon tetrachloride; disturbed heart action
Turpentine	Irritating to nose and throat; headache; vomiting; stomach pains
Xylene (xylol)	Same as for benzene except little damage to blood-forming organs

Extracted from "Useful Criteria in the Identification of Certain Occupational Health Hazards," Section of Industrial Hygiene of the Utah State Division of Health, Salt Lake City, Utah.

when spray painting in confined areas. A safety factor of 2 usually is acceptable for brush or roller application.

The volume of liberated solvent vapor, when applying paint containing 50 percent solvent at the rate of 10 gal per hr, is presented in Table 3-6. The fresh air requirement in order to maintain a vapor concentration of 25 percent LEL is given, as well as is the air requirement to just meet the TLV requirement.

Data presented in Table 3-6 clearly show the much greater ventilation requirements necessary to meet the TLVs for toxicity than are required to meet the lower explosive limit. Since ventilation equipment rarely is capable of meeting these large air demand requirements for toxicity, it is important that respirators with the proper chemical cartridge, or air-supplied respirators or hoods always be employed in confined locations. In addition, goggles also may be desired to protect eyes against irritation.

In ventilating tanks and other confined areas, it should be remembered that most solvent vapors are heavier than air and will settle to the tank bottoms. The fresh air inlet should be near the top of the confined

TABLE 3-6 Estimate of Fresh Air Requirement During Paint Application

(BASIS: 10 gal paint applied per hour; paint contains 50% solvent by volume)

Principal solvent	Vapor vol. per gal solvent, cu ft	Vapor vol. liberated, cu ft/hr	Air requirement, cfm	
			To meet 25% of LEL	To meet TLV
Acetone	41	205	530	3,400
Alcohol, ethyl	51	255	510	4,200
Alcohol, isopropyl	39	195	520	8,100
Benzene (benzol)	34	170	810	100,000
Carbon tetrachloride	31	155	0	260,000
Cyclohexane	27	135	690	7,500
Cyclohexanone	29	145	880	48,000
Cyclohexene	29	145	.....	8,000
Ethylene dichloride	38	190	200	63,000
Ethylene glycol ethyl ether	31	155	740	13,000
Ethylene glycol ethyl ether acetate	22	110	430	18,000
Ethylene glycol methyl ether	38	190	.....	130,000
Methyl ethyl ketone (MEK)	33	165	610	14,000
Methyl isobutyl ketone (MIBK)	24	120	670	20,000
Naphtha (coal tar)	30	150	1,100	12,000
Toluene (toluol)	29	145	800	12,000
Trichloroethylene	33	165	0	27,000
Turpentine	30	150	1,200	25,000
Xylene (xylol)	24	120	740	20,000

Note: Refer to Table 3-4, page 70, for trade names of the glycol ether solvents.

space, with one or more air discharge openings located near the bottom, so positioned as to eliminate dead air pockets.

The recommended ventilation when painting tank interiors is presented in Table 3-7. For 5,000-gal tanks and smaller, ventilation should provide 1½ air changes per min. For larger tanks, the location of the blowers is of great importance. Since solvent vapors are heavier than air, fresh air inlets should be located whenever possible near the top of the tank, and air exhaust fans should be positioned at one or more locations around the perimeter at the tank bottom.

On large storage tanks it is recommended to locate an exhaust blower diametrically opposite an air feed blower where shell plates have been removed at the tank bottom. Supplementary fans should be used to insure good air circulation with no dead air pockets in the tank. Ventilation should be continued after paint application until the paint is dry to touch.

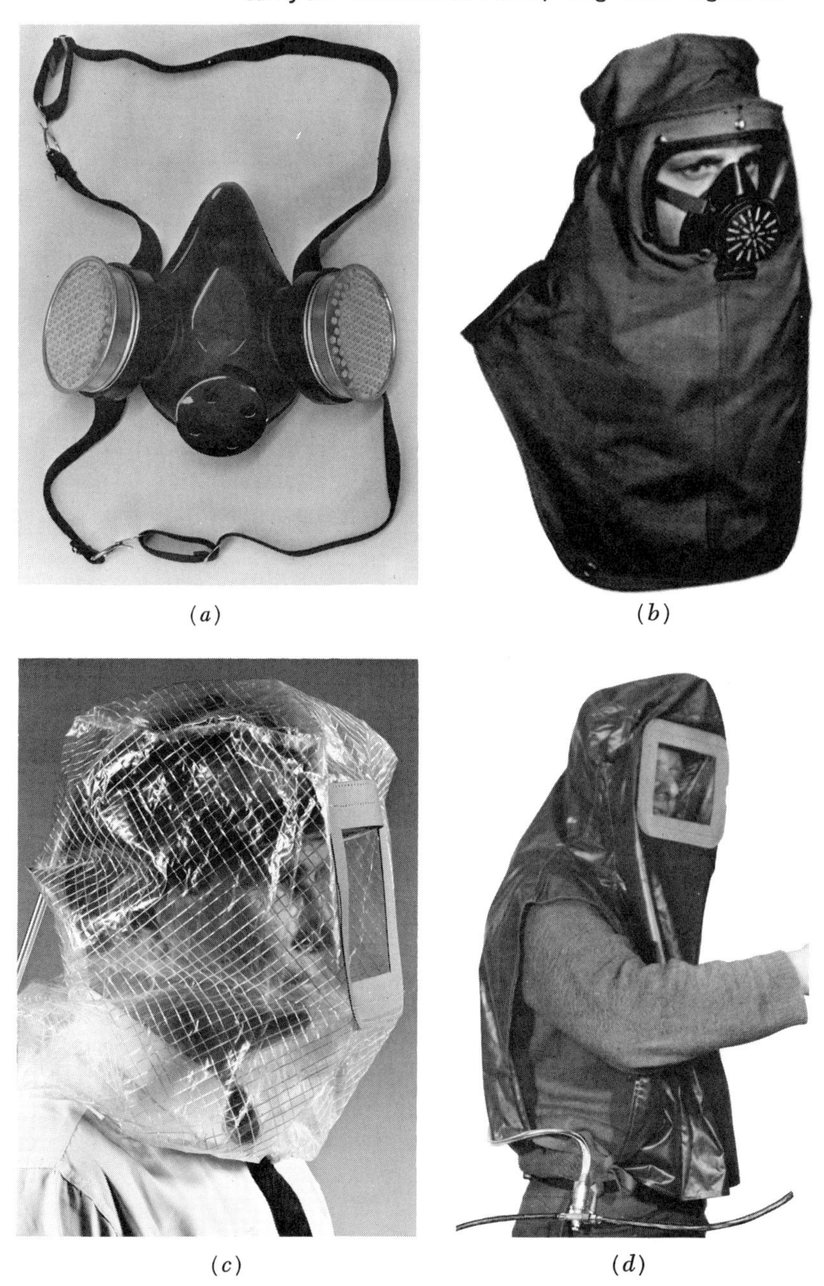

(*a*) (*b*)

(*c*) (*d*)

FIG. 3-7 Typical fume respirators and hoods. (*a, c, d—Binks Manufacturing Co., Chicago, Illinois.*) (*b—American Optical Co., Safety Products Division, Southbridge, Massachusetts.*)

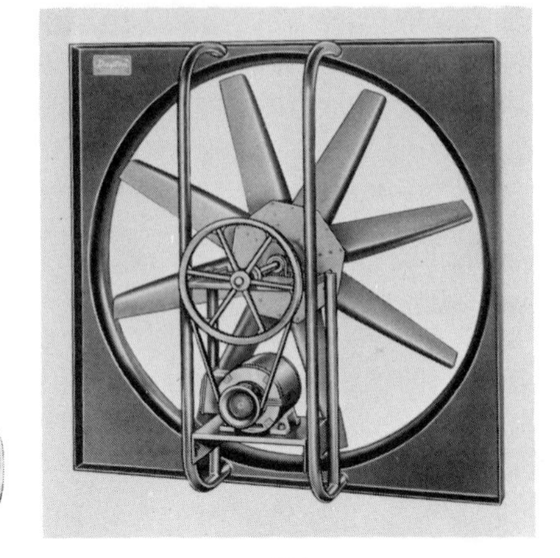

(*a*) (*b*)

FIG. 3-8 Examples of ventilators and air movers. (*a*) **Air-operated ventilator, capacity up to 5,000 cfm free air. Operating air requirement is 8 to 10 percent of air volume moved.** (*Mine Safety Appliances Co., Pittsburgh, Pennsylvania.*) (*b*) **Large volume air mover, capacity up to 70,000 cfm free air.** (*Dayton Electric Manufacturing Co., Chicago, Illinois.*)

The ventilation recommended in Table 3-7 is not sufficient to meet the Threshold Limit Values for safe breathing. During paint application, cartridge respirators or air-supplied hoods must be worn by the painters.

Personal protective equipment: Much of the following discussion is an extraction from "A Guide For Uniform Industrial Hygiene Codes or Regulations for Industrial Spray Coating," American Conference of Governmental Industrial Hygienists, September 1958, to which the reader is referred for additional information.

Workers engaged in spray-painting tank interiors or surfaces in other confined spaces shall be required to wear air-supplied respirators or hoods. In other locations where excessive rebound occurs or concentrations of dust particles or paint vapors are in excess of the Threshold Limit Values, either air-supplied or other types of respirators shall be used as appropriate. These other types include respirators with suitable chemical cartridges or dust filters. Illustrations of typical dust and fume respirators and hoods are presented in Figs. 3-6 and 3-7 on pages 65 and 73.

A filter shall be inserted in the air line to the respirator to remove oil mists, dusts, or other contaminants in the air system. The compressor shall be designed, and the air intake located, so as to prevent the entrance of carbon monoxide or other toxic gases, mists, or fumes into the system.

TABLE 3-7 Recommended Ventilation When Painting Tank Interiors

Air flow recommended to maintain solvent vapor concentrations below 10% of the lower explosive limit

Tank volume			Shell area, sq ft	Approx. air change rate, min/tank volume	Air mover capacity, cfm (free air)
Bbl	Gal	Cu ft			
119	5,000	670	280	0.6	1,000
238	10,000	1,340	450	1.0	1,200
356	15,000	2,000	680	1.3	1,500
500	21,000	2,800	850	1.4	2,000
1,000	42,000	5,600	1,130	2.0	2,500
1,500	63,000	8,400	1,415	3.0	3,000
2,000	84,000	11,200	1,860	3.0	4,000
2,500	......	14,000	2,360	3.0	5,000
5,000	......	28,000	3,770	4.5	6,000
10,000	......	56,000	5,340	5.5	10,000
15,000	......	84,000	7,240	5.5	15,000
20,000	......	112,000	7,540	5.5	20,000
30,000	......	168,000	10,100	5.5	30,000

All persons required to work in areas where their clothing is sprayed or splashed with paint or other coating materials or solvents should be furnished with protective clothing such as coveralls, aprons, or long sleeve impervious gloves. Protective covers should also be provided for shoes. Goggles, face shields, or other devices for protecting the face and eyes shall be furnished workers who might be sprayed or splashed with irritating solvents or coating materials (Fig. 3-9).

The use of protective ointments or barrier creams may be advantageous to prevent skin contact with paint pigments or other coating materials and various solvents. Their use may also facilitate the removal of paints from the exposed areas of skin. Caution should be exercised in choosing the appropriate barrier cream since only certain types will offer protection against paints and their respective thinners.

Raw solvents such as paint or lacquer thinners should not be used to remove paint or other coating materials from the skin. A number of satisfactory industrial cleaners, including waterless cleaning compounds, are available which remove these materials without harming the skin.

Frequent washing with soap or other industrial hand cleaners, especially before eating, is recommended to reduce the potential hazards of ingesting toxic substances and of skin disorders.

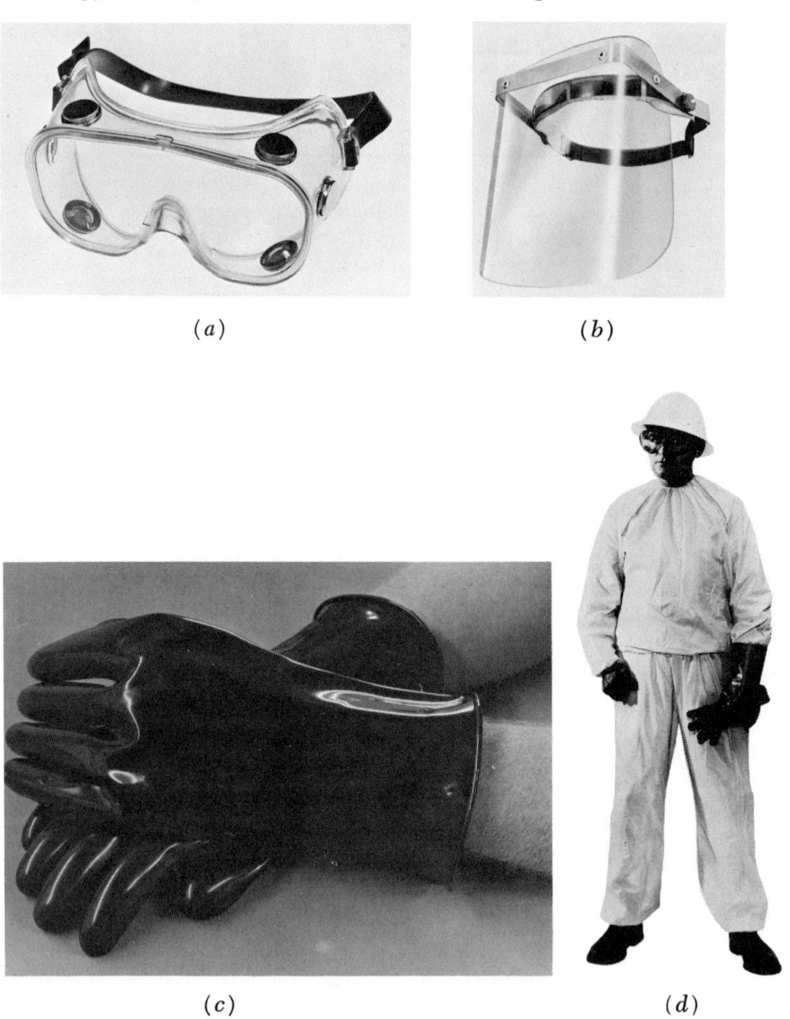

(a) (b) (c) (d)

FIG. 3-9 Protective clothing, goggles, and face shield. *(Binks Manufacturing Co., Chicago, Illinois.)*

Good safety habits should be established with respirators. They should be inspected regularly and expertly repaired as necessary, using only the proper parts. It is good practice to start each day with a clean respirator. With air-supplied respirators, check the air supply frequently for hose damage, loose connections, and proper location of the compressor air intake.

The recommended safety equipment and protective apparel to be used during surface preparation and paint application are summed up in Table 3-8 on page 77.

TABLE 3-8 Protective Equipment and Apparel to Be Used during Surface Preparation and Paint Application

Surface Preparation

Type of work performed	Hood, airline*		Respirator, chemical cartridge		Respirator, type F filter		Face shield†		Goggles, safety impact		Coveralls and hats		Gloves, leather		Gloves, rubber		Protective cream		Air movers‡	
	O	C	O	C	O	C	O	C	O	C	O	C	O	C	O	C	O	C	O	C
Wire brushing, chipping, scraping, and grinding					x	x	x	x	x	x	x	x	x	x						
Sandblasting:																				
Operator	x	x									x	x	x	x						
Other workmen in the vicinity		x			x		x		x		x	x	x	x						
Paint removing				x			x	x			x	x			x	x	x	x		x
Solvent cleaning				x			x	x			x	x			x	x	x	x		x

Paint application

Type of work performed	Hood, airline*				Respirator, chemical cartridge				Respirator, type F filter				Face shield†				Goggles, safety impact				Coveralls and hats		Gloves, leather		Gloves, rubber		Protective cream		Air movers‡	
	Brush		Spray		Brush		Spray		Brush		Spray		Brush		Spray		Brush		Spray		Brush or spray		Brush or spray		Brush or spray		Brush or spray		Brush or spray	
	O	C	O	C	O	C	O	C	O	C	O	C	O	C	O	C	O	C	O	C	O	C	O	C	O	C	O	C	O	C
Hot-applied coal-tar and asphalt coatings				x		x							x	x							x	x	x	x			x	x		x
Cold-applied coal tar coatings						x		x													x	x	x	x			x	x		x
Coal-tar epoxy coatings				x		x							x	x	x						x	x	x	x			x	x		x
Alkyd-type coatings						x		x													x	x	x	x			x	x		x
Phenolic oil varnish coatings						x		x													x	x	x	x			x	x		x
Catalytically cured phenolic and phenolic epoxy coatings				x		x							x	x	x						x	x	x	x			x	x		x
Vinyl coatings		x		x	x		x						x		x						x	x	x	x			x	x		x
Wash primer																					x	x	x	x			x	x		x
Inorganic zinc coating														x							x	x			x	x				
Chlorinated rubber paints						x		x													x	x	x	x			x	x		x
Lacquers		x		x	x		x						x		x						x	x	x	x			x	x		x
Water-base paints																					x	x			x	x	x	x		
Shellac						x		x					x	x							x	x	x	x			x	x		x

* Airline hoods shall always be used in confined spaces having a volume of less than 5,000 gallons; and as indicated.
† Face shields shall always be used when working overhead; and as indicated.
‡ Air movers shall be used in confined spaces as indicated, except where sufficient ventilation is provided through open windows and doors.
§ Key: O = Open areas; C = Confined areas.

Smog Control Legislation

Smog control legislation enacted by San Francisco and Los Angeles, concerned with liberation of photochemically reactive organic solvents to the atmosphere, is of far-reaching importance. Their impact on paint and solvent formulation is essentially nationwide, and consequently a brief discussion is warranted in this manual.

Both legislations control the quantity of reactive solvents that may be emitted, and differ mainly in methods of enforcement. Photochemically reactive solvents, unfortunately, include the commonly employed strong solvents such as toluene, xylene, methyl isobutyl ketone, and trichloroethylene. As a result of these restrictions, most major paint formulators throughout the country are now active in one or both of two programs: (1) formulating durable, weather-resistant, and high gloss water-base paints and coatings; and (2) developing acceptable solvent substitutes for the reactive solvents which have been employed in their paint formulations.

The principal solvent properties of importance in paint formulation and application are solvent power and evaporation rate. Blends of nonreactive solvents which satisfactorily match these two properties for most, if not all, paint solvents may be formulated from properly selected aliphatic and naphthenic hydrocarbons and their alcohol, ester, and ether derivatives. Examples of these acceptable derivatives are:

Esters	Alcohols	Ethers
Ethyl acetate	Isopropyl alcohol	Ethylene glycol ethyl ether
n-Propyl acetate	Isobutyl alcohol	Ethylene glycol methyl ether
Isobutyl acetate	*n*-Butyl alcohol	Diethylene glycol ethyl ether
Methyl amyl acetate	Methyl amyl alcohol	Diethylene glycol methyl ether
Ethylene glycol ethyl ether acetate	*n*-Hexyl alcohol	Diethylene glycol butyl ether
Diethylene glycol ethyl ether acetate	*n*-Octyl alcohol	
Glycol diacetate		

Note: Refer to Table 3-4, page 70, for trade names of the glycol ether solvents.

The commonly employed solvent mineral spirits is a petroleum distillation cut between 300 and 400°F. Mineral spirits consists of blends of aliphatic, naphthenic, and aromatic hydrocarbons, the exact composition depending upon the crude oil source. Without adjustment, it usually contains from 10 to 25 percent aromatics. Since the aromatic content now is limited to 8 percent for use in Los Angeles, a loss in solvent power would result in this adjustment unless compensated by an increase in the naphthenic content of the spirits.

A brief review of the more important features of both Regulation 3 (San Francisco) and Rule 66 (Los Angeles) concludes this chapter.

Regulation 3—San Francisco: Regulation 3 was developed by San Francisco authorities to control the emission of photochemically reactive organic compounds from industrial sources. Regulation 3 defines these reactive compounds as:

1. All olefins (organic compounds containing the —C=C— bond), except for benzene and the halogen-substituted benzenes
2. All substituted aromatics (organic compounds containing the benzene ring) such as toluene and xylene
3. All aldehydes and ketones (organic compounds containing the =C=O group) such as MEK and MIBK

Industrial sources of emission include the manufacture and use of paints, coatings, and adhesives; petroleum refining and marketing; coating lines; drum and oven driers; and other users of organic solvents or solvent-containing materials.

The basic regulation states that no source can emit a total concentration of all organic compounds greater than 50 ppm (calculated as hexane), unless:

1. Less than 20 lb per day total, or 10 lb per day reactive, organic compounds are emitted; *or*
2. The organic compounds emitted contain less than 5 percent by volume of reactive compounds; *or*
3. The use of control (abatement) equipment results in a reduction of 85 percent of the reactive organic compounds which otherwise would be discharged to the atmosphere.

The regulation applies in its *least* restrictive form. The concentration of organic compounds is calculated in terms of hexane since combustible-gas detectors commonly employed for these determinations usually are calibrated against hexane.

Special provisions apply to the use of paints. According to Regulation 3, no person shall sell or offer for sale to a buyer for use, in the San Francisco area, noncomplying paints or solvents in 1-qt containers or larger, except:

1. For sale to an industrial user, *or*
2. For sale to a formulator who will add complying solvents until the blend complies.

Complying solvents and coatings are defined in Regulation 3 as follows:

1. The volatile portion of complying solvents and surface coatings may contain up to 8 percent by volume reactive organic solvents and in addition up to 12 percent by volume monosubstituted aromatic solvents (such as toluene);
2. Complying industrial coatings may contain up to 20 percent by volume reactive organic compounds in their volatile fraction. In water-containing paints, the water content is included in these determinations.

Rule 66—Los Angeles: Rule 66 was developed by Los Angeles authorities to reduce the emission of photochemically reactive volatile liquids into the atmosphere. Photochemically reactive liquids and solvents are defined in Rule 66 as:

1. Any liquid, including water, which contains more than a 20 percent aggregate total of:
   a. All olefinic or cycloolefinic compounds, *and*
   b. All substituted aromatic hydrocarbons, *and*
   c. The branch chain ketones, and trichloroethylene
2. Any liquid, including water, which contains more than the indicated percentage concentrations of combinations of the following groups of solvents:
   a. 5 percent total blends of olefinic or cycloolefinic compounds, *or*
   b. 8 percent of blends of aromatic hydrocarbons (except toluene and ethyl benzene), *or*
   c. 20 percent total of blends of toluene, ethyl benzene, branch chain ketones, and trichloroethylene

Examples of olefinic solvents (see *a* above) employed in paint formulation are isophorone, mesityl oxide, and dipentine; examples of aromatic hydrocarbons (see *b* above) are xylene, tetralin, and cumene. Exempt solvents encompass the aliphatic and naphthenic compounds including their halogenated derivatives, and also include perchloroethylene.

The basic provision of Rule 66 states that no one source may emit daily, at atmospheric temperature, more than 40 lb of organic liquids containing photochemically reactive compounds; or at elevated temperatures (oven dry), more than 15 lb of all types of organic volatile materials, unless the use of air pollution control equipment has reduced the total emission which otherwise would occur by 85 percent.

Special provisions apply to the application of paints, as covered in Rule 66.1, "Architectural Coatings." Rule 66.1 prohibits the sale or application of an architectural coating containing photochemically reactive solvents as defined in Rule 66 (above) when in one quart or larger containers, nor shall photochemically reactive solvents be employed for thinning purposes.

Multicomponent coatings are considered in their combined form in determining compliance with Rule 66.1. Architectural paints are defined as those applied to residential, commercial, or industrial buildings or their appurtenances.

Chapter **4**

Surface Preparation and Pretreatment

Introduction

This chapter discusses the more important surface-preparation and pretreatment methods most commonly employed by contractors and others, for the field application of paints and protective coatings. In addition, several important shop surface-preparation methods are briefly reviewed.

The first and certainly most important step in good painting practice is to be sure that the surface has been properly prepared for the paint system to be applied. The importance of proper surface preparation cannot be overemphasized. The investment made in the purchase and application of the best of coatings will be almost entirely lost if the surface preparation is inadequate. All too often the premature failure of a recommended coating system is blamed on the manufacturer of the material, when actually the surface preparation was skimped as being an unnecessarily expensive luxury.

Surface-preparation methods vary both in cost and effectiveness. Since the minimum preparation requirement differs for coatings exposed to the various service conditions, it is economically important to select the optimum method and degree of surface preparation consistent with the service condition and coating to be applied.

Table 4-1 presents a brief summary of the recommended surface preparation requirements for various service conditions. It generally is considered a good investment to insure that the quality of surface preparation for coatings for atmospheric exposure is one grade better than the minimum specified in Table 4-1. For immersion service, most coatings require only the best in surface preparation anyway.

The effect of surface preparation on paint-service life is illustrated by two exposure test results reported by the Steel Structures Painting Council:

1. In a test conducted in England, the same paint system (an oleoresinous paint for exterior exposure) exhibited a considerably longer service life when applied to better-prepared surfaces:

Surface preparation method	Two primer, two topcoats	No primer, two topcoats
Weathered, wire brushed....	2.3 years	1.2 years
Pickled..................	9.5+ years	4.6 years
Sandblasted..............	10.4 years	6.3+ years

2. In an extensive panel exposure test, the relative effectiveness of commonly employed surface-preparation methods for several oleo-

TABLE 4-1 Mechanical Surface Preparation Recommended for Various Service Conditions

Exposure condition	Mechanical surface preparation	Minimum required for:
Immersion: chemicals; acids; salts; distilled water	White-metal blast	Zinc silicates, vinyls, phenolics, silicones, chlorinated rubbers
Immersion: water; brine; oils *and* *Atmosphere:* chemical	Near-white blast	Organic zinc-rich, vinyls, phenolic varnishes, epoxies, coal-tar epoxies, chlorinated rubbers
Atmosphere: industrial; humid; marine	Commercial blast	Coal-tar epoxies, epoxy esters, phenolic varnishes, chlorinated rubbers
Atmosphere: uncontaminated; humid	Brush blast	Oil-base, alkyds, epoxy esters, bituminous, coal-tar epoxies
Atmosphere: uncontaminated; exterior and interior	Power tool and hand tool	Oil-base, water-base, alkyds, bituminous

Comments on Other Surface Preparation Methods:

1. *Solvent and chemical cleaning:* Employed for special contaminations, coatings, or service conditions. Quality of preparation varies widely.
2. *Flame cleaning:* May be substituted for brush blast cleaning for most paints satisfied by this surface preparation quality.
3. *Pickling and phosphatizing:* May be substituted for white-metal, near-white-metal, or commercial blast, depending upon quality and service condition.

resinous paints in atmospheric exposure was determined and summarized as follows:

Surface preparation method	*Effectiveness rating*
Pickled and phosphatized	93%
Sandblasted	83%
Flame cleaned (3 passes)	73%
Flame cleaned (1 pass)	47%
Pickled only	53%
Hot phosphatized only	40%

Paint holds to the metal surface by two mechanisms: chemical attraction or "adhesion," and mechanical anchoring or "bonding." Consequently, the desired surface preparation should result in a clean surface which has been slightly roughened. Roughening provides a "tooth" or anchor for mechanical bonding, and also increases the surface area for increased adhesion often by as much as two to three times.

Effective surface preparation starts with a preliminary degreasing to remove any oil or grease which is present in appreciable quantities, since these are not effectively removed by mechanical methods other than sandblasting. After surface preparation by mechanical methods, a surface pretreatment before painting may be recommended under certain conditions to wet or etch the surface, or to provide for increased adhesion of the primer.

The following discussion of each surface-preparation method presents a brief review of the most important techniques and equipment. For more detailed information reference should be made to the excellent publications of such organizations as the Steel Structures Painting Council (Pittsburgh, Pennsylvania), the National Association of Corrosion Engineers (Houston, Texas), or to the manufacturers' instruction manuals.

Also included for immediate reference purposes are the principal or special safety and health hazards associated with each of these methods. Chapter 3 of this manual should be reviewed for additional and more general safety considerations which apply to surface-preparation hazards.

Hand-tool Cleaning

Hand cleaning of steel may be defined as the removal of all loose mill scale, loose rust scale and dirt, and nonadherent old paint by means of hand tools. Hand cleaning does not remove tight mill scale, paint, and other tightly adhering contaminants, or all traces of rust, or deposits in pits and crevasses.

This is a slow operation, and is recommended only for limited areas such as in preparation for touch-up painting. Hand cleaning is suitable only for the application of bituminous coatings, and oleoresinous and water-base paints for atmospheric exposure. Furthermore, the use of a wetting oil pretreatment is recommended when applying an oleoresinous paint over a hand-cleaned surface.

Hand cleaning, on the other hand, often is used to remove heavy deposits of scale and paint before employing more thorough surface-preparation methods such as power-tool cleaning and sandblasting. Hand scraping and wire brushing almost always are a part of the flame-cleaning process to complete the removal of flame-loosened deposits.

Commonly employed tools are chipping hammers, rust hammers, chisels, scrapers, putty knives, wire brushes, and sometimes emery cloth or sandpaper. Auxiliary equipment includes dust brushes, brooms, and safety equipment as required.

Before hand cleaning is undertaken, the surface should be examined to determine the amount and nature of the contaminants. If detrimental amounts of oil or grease are present, solvent cleaning must precede hand cleaning.

The chipping hammer and sharp-pointed rust hammer are used to loosen and remove nonadherent scale by impact. Heavy scale is removed by chisels; loose voluminous rust is easily removed by scrapers or knife blades. Additional rust is removed by wire brush or emery cloth.

Care should be taken to avoid spreading any contaminants such as gobs of grease over the entire surface through the cleaning operation. Also, avoid making deep gouges into the metal from impact tools. Sharp burrs on the metal surface will shorten the effective life of the coating. Also avoid excessive polishing and burnishing of the metal, which reduces coating adhesion.

Wire brushes which are no longer effective because of lost or badly bent bristles should be discarded. Chisels and scrapers should be kept sharp to be effective.

After cleaning is completed, the surface must be brushed, swept, dusted, or blown off with compressed air to remove all loose matter. If detrimental amounts of grease or oil are still present, these areas shall be spot cleaned with solvent.

The degree of cleanliness after hand cleaning depends upon the initial condition of the steel. It is obvious that greater amounts of rust and scale will remain on badly pitted steel than on a smooth surface.

The Steel Structures Painting Council has established an arbitrary standard for hand cleaning (SP-2-63), as being that surface cleanliness which results from vigorous hand brushing with a new wire brush at a rate of 2 sq ft per min after all stratified rust scale previously had been removed.

Nonsparking tools must be employed where flammable vapors may

FIG. 4-1 Hand-cleaning tools.

be present. Operators engaged in surface preparation shall always wear safety goggles, gloves, and a safety helmet, and a dust respirator if dust is generated. If the surface was previously coated with a lead- or zinc-base primer or with an antifouling marine paint, a dust respirator must be worn.

Power-tool Cleaning

Power-tool cleaning of steel may be defined as the removal of all loose mill scale, rust scale, nonadherent paint, dried or caked dirt, and most of the superficial rust by means of power-operated tools. Power-tool cleaning does not remove tight mill scale, nor all traces of rust on pitted steel. Power-tool cleaning is not recommended for badly pitted surfaces.

Contaminations of an oily or greasy nature are not removed by power-tool cleaning, but in fact are spread over the entire surface. It is important that these be removed first by proper solvent wiping.

Without care, power brushing will leave a polished surface through the action of rust (rouge) on the surface. The desired anchor pattern for paint adhesion is not formed by power-tool cleaning.

Power-tool cleaning is employed on steel structures where blast cleaning is impractical or uneconomical, and the coating systems used are those which tolerate the contaminants left behind after such cleaning. These are the bituminous coatings and oleoresinous and water-base paints for atmospheric exposure. As with hand cleaning, pretreatment of the surface with a wetting oil is desired before applying an oleoresinous paint to a power-tool-cleaned surface.

Power-driven tools include pneumatic chippers, chisels, descaling tools and needle hammers, rotary scalers, rotary wire brushes, and abrasive wheels. Auxiliary equipment includes the air or electric power supply, dust brushes, brooms, and safety equipment as required.

The tools must be kept sharp, otherwise they have a tendency to drive rust and scale into the surface. Great care must be exercised in using these tools so as not to cut into the surface excessively, removing sound metal and leaving sharp burrs where the paint will fail prematurely.

Heavy deposits of rust scale, loose mill scale, thick old paint, weld flux, slag, and other brittle products are removed by rotary or piston-type impact tools. Additional rust is removed by rotary wire brushes of various shapes, or by abrasive wheels. Grinding is used to smooth weld seams and to round sharp edges and corners. All rivet heads, corners, joints, and openings will need special attention to be cleaned properly.

After cleaning is completed, the surface is brushed, swept, or blown off with compressed air to remove all loose matter. If detrimental amounts of grease or oil are still present, these areas shall be spot cleaned with solvent.

The degree of cleanliness after power-tool cleaning depends upon the original condition of the surface. An arbitrary standard for cleanliness has been established by the Steel Structures Painting Council

(*a*) **Grinding wheels and brushes**

FIG. 4-2 Power-cleaning tools. (*Chicago Pneumatic Tool Co., New York, New York.*)

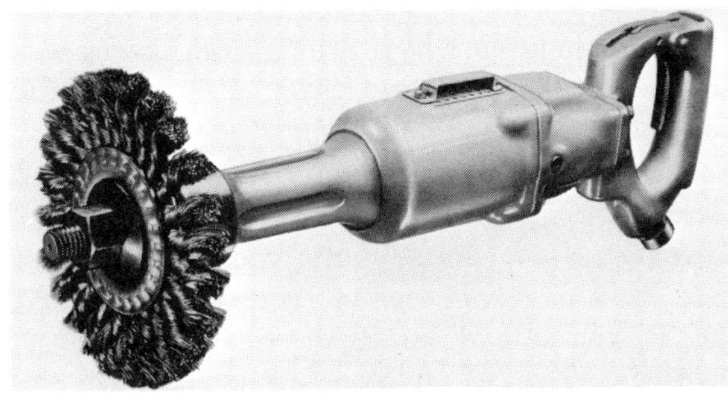

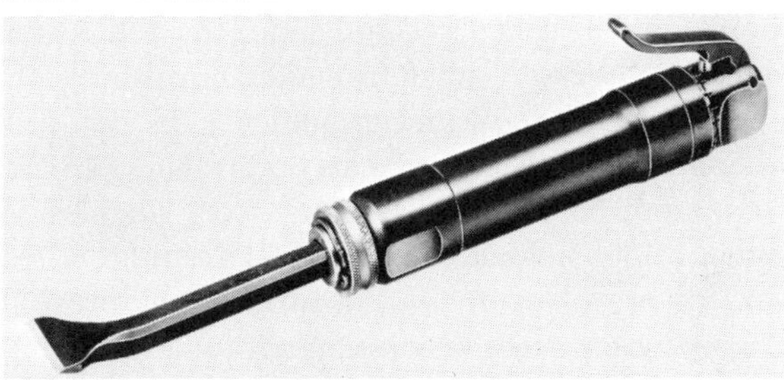

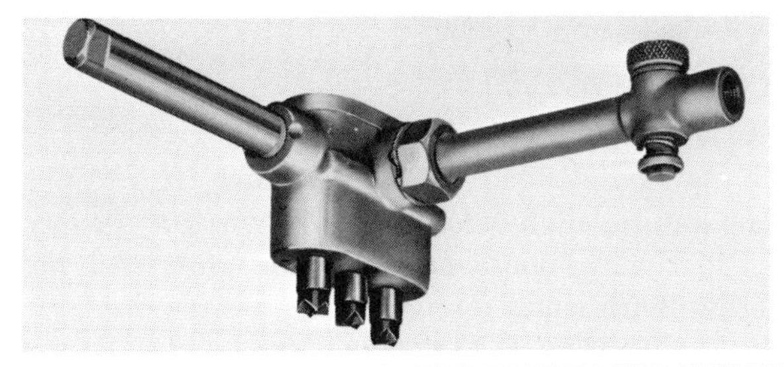

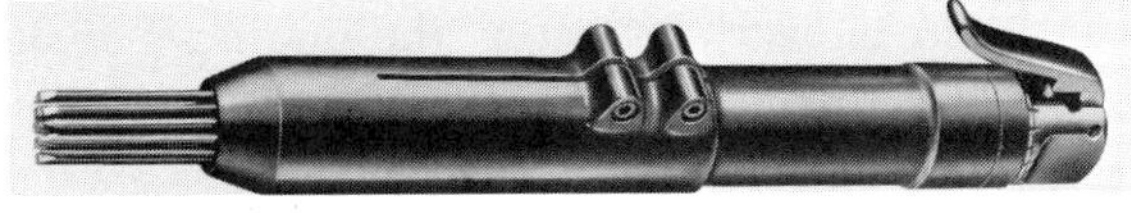

(b) Grinders, chippers, and scalers

(c) Grinder in use

FIG. 4-2 (Continued)

(SP-3-63) as being the surface condition which results from power wire brushing at a cleaning rate of 2 sq ft per min, after all stratified rust scale previously had been removed. In establishing this standard surface condition, a commercial wire-brushing machine operated at a speed underload of 3,450 rpm and equipped with a 6-in.-diameter cup brush of double-row knotted construction No. 20-gauge music wire shall be used. It is clear that power wire brushing will remove more tight rust than vigorous hand brushing at the same rate of 2 sq ft per min.

Safety considerations govern the speed of rotary tools; do not exceed the maximum speed recommended by the manufacturer. Furthermore, high speeds result in detrimental burnishing of the surface.

Nonsparking tools and grounded explosion-proof motors must be used where flammable vapors may be present. Operators engaged in surface preparation shall always wear safety goggles or a face shield, gloves, and a safety helmet. Air or dust respirators shall be worn where dust hazards are present. If the surface was previously coated with a lead- or zinc-base primer or with an antifouling marine paint, a dust respirator must be worn.

Flame Cleaning

Flame cleaning of steel may be defined as the cleaning that results from passing a high-temperature, high-velocity flame over the surface and then wire brushing to remove loosened scale and rust. All unbonded scale, rust, and other detrimental foreign matter are removed.

This method of surface preparation is not widely used today except in certain industries such as the marine industry. Flame cleaning can be used only in nonhazardous locations. It results in a surface condition equivalent to brush blasting, but with the added benefit of a warm and dry surface. Paints should be applied while the temperature of the steel is still well above that of ambient air.

Paint removal by flame may be more expedient and economical than by use of chemical paint removers. Old oil-base paints are removed easily by scraping after flame treatment. Most protective coatings, on the other hand, must be burned to carbon and these carbonized remains are difficult to remove.

The successful flame-cleaning process depends upon the rapid heating of scale and deposits down to the metal surface, to a temperature around 300°F. A high velocity, high temperature clean flame is required, such as the oxyacetylene flame. The flame is passed over the surface at a 45 to 60° angle, at an optimum speed which will be somewhere between 15 and 25 ft per min.

The desired speed of traverse shall be such that dirt, rust scale, unbound mill scale, and similar foreign matter are freed by the rapid, intense heating of the flames. The surface of red granulated rust is reduced to a black powder. The speed shall not be so slow that the scale is fused to the surface of the steel or that thin sections of steel are warped, but shall be slow enough to evaporate all moisture which condenses on the steel from the gas-combustion products.

The flame-cleaning equipment consists of a blow pipe similar to the blow pipe used in oxyacetylene torches, with an extension tube to provide the desired length, and a flame-cleaning head having a row of gas ports about ⅛ in. apart. Heads are available in various sizes and shapes to suit the size and shape of the structural steel to be cleaned.

Gas hoses, pressure regulators, and gas supplies are the same as for conventional oxyacetylene equipment. Gas cylinders are set up in a safe location and usually are connected by a common manifold to maintain a uniform and adequate flow of gas.

Deleterious amounts of oil and grease should be removed well in advance of the flame-cleaning operation. Welding flux and spatter also must be removed before flame cleaning.

Promptly after application of the flames the surface must be wire brushed, and hand scraped whenever necessary, and then swept or dusted to remove all loose material.

The steel shall be free of all rust, loose mill scale, or other foreign matter and shall be completely dry and well above the ambient temperature at the time the prime coat is applied. The warm surface results in a better flow-out and penetration of pores and crevasses by the first coat applied to the surface.

One torch operator with a 6-in. head will serve three or four wire brushers and one or two painters. He should keep from ½ to 1 hr ahead of painting. A surface with a moderate amount of rust and loose scale will be cleaned at a rate of 300 to 400 sq ft per hr; a heavily

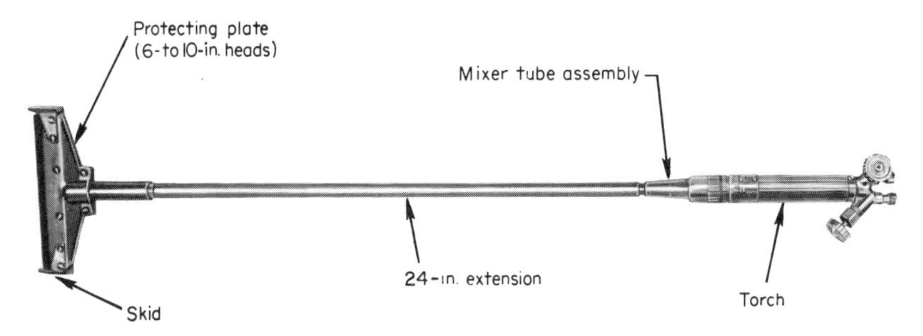

FIG. 4-3 Flame-cleaning torch. *(Union Carbide Corp., Linde Division, New York, New York.)*

rusted surface, with angles, welds, and rivets, would be cleaned and painted at a much slower rate.

Safety: Flame cleaning and priming operations shall be so scheduled and conducted as to be safe. With the open flame present in flame cleaning, special care must be exercised in locating all paint and solvent supplies, and especially in determining that no flammable gases are present.

Unless priming can follow the flame cleaning at a safe distance with adequate ventilation, the flame cleaning must be interrupted while priming cleaned areas. Refer also to Chap. 3 of this manual where fire hazards are discussed in more detail.

Operators engaged in surface preparation shall always wear safety goggles, gloves, and a safety helmet. Adequate ventilation shall be provided to remove any toxic fumes that may be generated by heating the old paint and surface contaminants.

Abrasive Blast Cleaning

Introduction: Abrasive blast cleaning may be defined as cleaning through the impact of abrasive particles propelled at high velocity against the surface to be cleaned.

Abrasive blast cleaning is one of the few methods which effectively removes all traces of oil or grease from the surface. In addition, blast cleaning produces a very desired roughening of the surface termed "anchor pattern," which aids in the adhesion of paint.

Two principal methods are employed for propelling the abrasive particles. These are (1) through the centrifugal action of electrically driven wheels or paddles to physically throw the particles against the surface, and (2) by means of the jet action of compressed air as it escapes from a restricting orifice. The first method, "wheelblast," is limited to shop cleaning because of the size and complexity of the installation. A brief description of this cleaning process is presented at the end of this section.

Air blast cleaning is the most commonly employed field method of surface preparation. It is convenient, portable, fast, and suitable for cleaning any shape to any desired degree of cleanliness. Several variations of air blast cleaning are dry and wet sandblasting and vacuum blasting. Since sand is the most commonly employed abrasive, the term "sandblasting" is generally applied to all air-blast cleaning regardless of the abrasive used.

Standards of surface preparation: The degree of surface cleanliness obtainable through abrasive blast cleaning may be varied over a wide range, as desired. The Steel Structures Painting Council has developed four standards of surface preparation which are accepted and recognized by professional organizations and industry. These are:

1. SSPC-SP 5-63 White-metal Blast Cleaning: This surface finish exhibits a gray-white uniform metallic color, the exact shade of whiteness depending upon the abrasive employed. The surface is entirely free of oil, grease, dirt, visible mill scale, rust, paint, and any other deposit. The surface is slightly roughened to form a suitable anchor pattern for coatings.

2. SSPC-SP 10-63T Near-white Blast Cleaning: This surface finish exhibits very slight streaks or slight discolorations caused by rust stain, mill scale, or minute tight residues of paint that may remain. Except for this discoloration, the surface is entirely free of all oil, grease, dirt, mill scale, rust, paint, and any other deposit. The surface is roughened to form a suitable anchor pattern.

3. SSPC-SP 6-63 Commercial Blast Cleaning: This surface finish exhibits slight streaks and discolored areas due to rust stain, mill scale, or slight tight residues of paint that may remain. The surface is entirely free of all oil, grease, dirt, loose rust, and deposits; and mill scale, rust scale, and oil paint are removed except for the small amount remaining to cause the discoloration described. If the surface is pitted, slight residues of rust or paint may be found in pit bottoms. At least two-thirds of each square inch shall be free of all visible residues, and the remaining area limited to the light discoloration described above.

4. SSPC-SP 7-63 Brush-off Blast Cleaning: This surface finish is entirely free of all oil, grease, dirt, loose rust, loose mill scale and loose old paint, but tight mill scale and tightly adhering rust scale and paint are permitted to remain. However, all remaining mill scale and rust are to be exposed to the abrasive blast sufficiently to expose numerous flecks of the underlying metal over the entire scale surface.

It is obvious that the more thorough the surface preparation, the longer it will take and the more costly it will be. Consequently the white-metal and near-white-metal blast finishes usually are specified only for coatings in immersion service, while commercial and brush-off finishes are suitable for most atmospheric exposure conditions. (Refer to Table 4-1, page 84).

Approximate relative cleaning rates for the different degrees of surface finish on moderately rusted and pitted steel would be:

White-metal Blast:	80 to 120 sq ft per hr
Near-white Blast:	120 to 150 sq ft per hr
Commercial Blast:	200 to 250 sq ft per hr
Brush-off Blast:	400+ sq ft per hr

These rates are based on use of a ⅜-in. venturi nozzle, with silica sand driven by air at 100-psig nozzle pressure.

Blast-cleaning abrasives: Sand is the most commonly used abrasive for air blast cleaning in the field, since it is reasonable in cost and almost always locally available. Its breakdown rate is fairly high, however—between 10 and 40 percent—and so it is seldom reclaimed for reuse. Sand with a 90 percent or higher silica content is desired for more effective cleaning and lower breakdown rates. Sand with breakdown rates above 20 percent should not be used for sandblasting.

The different abrasives in common use for surface preparation for painting are:

Silica Sand—Shop and Field Use: Cheap; effective; available; not reused; generates dust; hazard of silicosis.

Other natural abrasives (crushed flint, garnet sand)—Shop and Field Use: Efficient; quicker cutting; less dust; lower breakdown rate; usually reclaimed and reused.

Slag (by-product of lead and copper ore reduction)—Shop and Field Use: Efficient; fast cutting; less dust and no silicosis hazard; high breakdown rate so not reused.

Nonmetallics (silicon carbide, aluminum oxide)—Shop Use: Not subject to rust; low dust; very hard; fast cutting; costly; reclaimed and reused.

Metals (cast iron and steel shot and grit)—Shop Use: Used only indoors due to rusting; dust-free; low breakdown rate; efficient; costly; reused.

Abrasive shapes may be classified as:

Semisharp: The common shape of sand and slag.
Grit: Angular; gouges the surface; high cutting efficiency.
Shot: Spherical; peens the surface; pounds off brittle deposits and mill scale; may pound impurities into the surface.

Grit and shot shapes usually apply to the metal abrasives. Figure 4-4 illustrates the type of surface preparation which is produced by their use.

Contrary to what one may think, the cleaning rate of an abrasive increases with smaller mesh size, until a minimum size is reached at which cushioning occurs. A mesh size range of 20 to 50 (U.S. Sieve Series) now is often used, whereas coarser particle sizes had been employed formerly. Finer particles—100 mesh and finer—are not recommended because the desired anchor pattern is not formed on the surface, while particles larger than 16 or 18 mesh may gouge the surface too deeply.

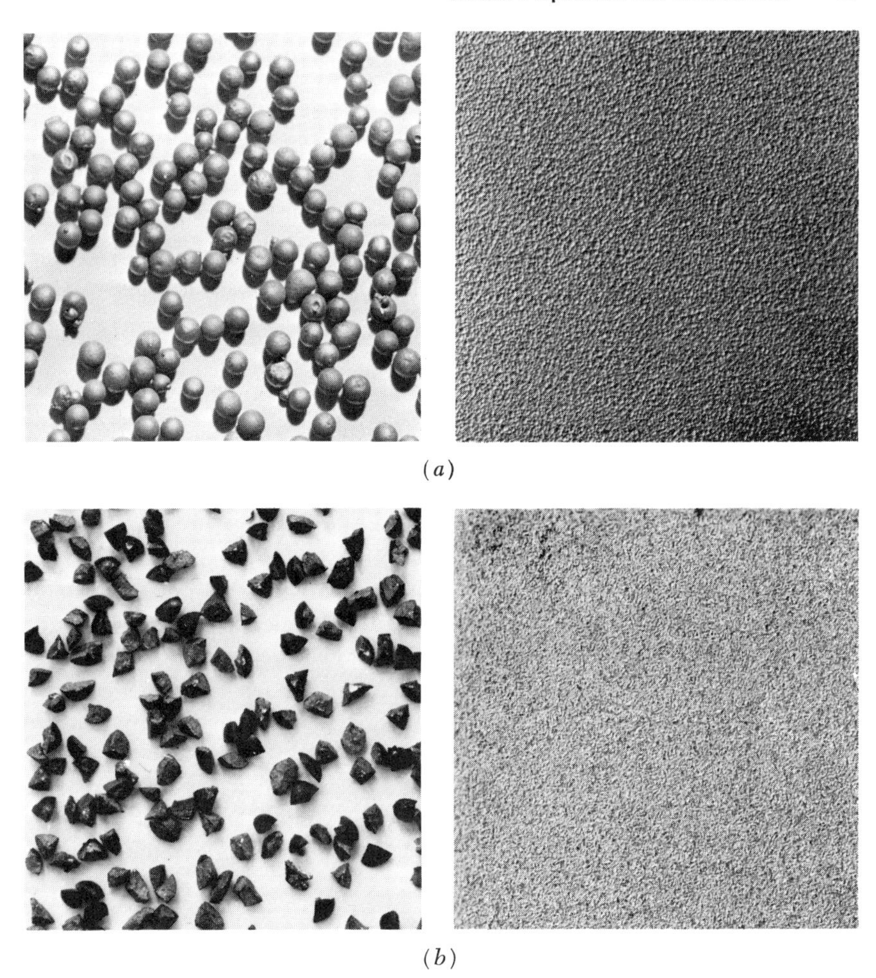

(*a*)

(*b*)

FIG. 4-4 Anchor patterns from shot and grit. (*a*) **Shot particles and shotblast surface;** (*b*) **grit particles and gritblast surface.** (*Pangborn Division, The Carborundum Co., Hagerstown, Maryland.*)

For most paints and coatings, the desired anchor pattern is from $1\frac{1}{2}$ to $2\frac{1}{2}$ mils, which is the profile height from pit bottom to tip of peak. Heavier coatings such as the mastics, coal-tar epoxies, etc., can be applied over heavier anchor patterns. Anchor patterns are easily measured by grinding a flat on the surface down until the pits just disappear, and determining the depth by resting a depth micrometer or dial indicator on the sand-blasted surface. Magnetic instruments for measuring anchor-pattern depth, such as shown in Fig. 4-5, also are available and convenient to use.

Representative anchor patterns for a number of different types and

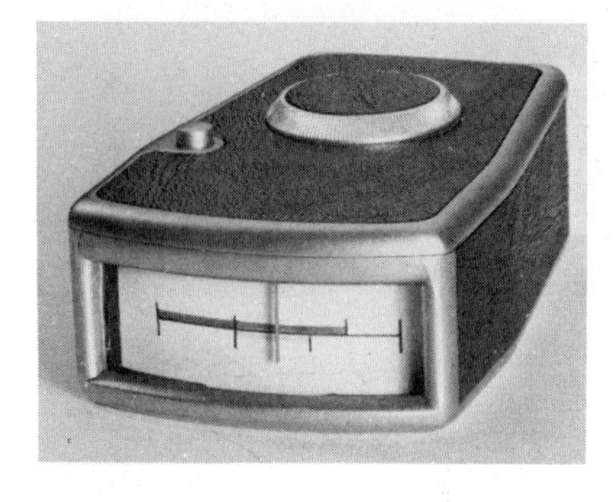

FIG. 4-5 Magnetic anchor pattern depth gauge: Elcometer Roughness Gauge. *(Gardner Laboratory, Inc., Bethesda, Maryland.)*

sizes of abrasives, as determined in actual blast cleaning operations, are presented in Table 4-2.

Since the life of protective coatings in most services depends upon the coating thickness which exists on top of the peaks, it is not enough to specify the paint thickness to be applied without taking the anchor pattern into account. It is becoming present practice to request, instead, "anchor pattern plus x mils." Knowledge of the anchor pattern produced therefore is becoming of increasing importance in specifying the surface preparation for a coating system for a corrosive service.

Sandblast equipment: Several types of dry sandblast equipment are in common shop and field use. These are:

Vacuum Blast: The blast nozzle is enclosed in a vacuum cup, and the abrasive is reclaimed from the immediate blast area by vacuum. Capacity is limited to moderate areas; most useful

TABLE 4-2 Anchor Patterns by Different Abrasives

Abrasive	Maximum particle size	Maximum profile ht.
Sand, very fine	Through 80 mesh	1.5 mils
Sand, fine	Through 40 mesh	1.9 mils
Sand, medium	Through 18 mesh	2.5 mils
Sand, coarse	Through 12 mesh	2.8 mils
Iron grit G-50	Through 25 mesh	3.3 mils
Iron grit G-40	Through 18 mesh	3.6 mils
Iron grit G-25	Through 16 mesh	4.0 mils
Iron grit G-16	Through 12 mesh	8.0 mils
Iron shot S230	Through 18 mesh	3.0 mils
Iron shot S330	Through 16 mesh	3.3 mils
Iron shot S390	Through 14 mesh	3.6 mils

SOURCE: Steel Structures Painting Council, Pittsburgh, Pa.

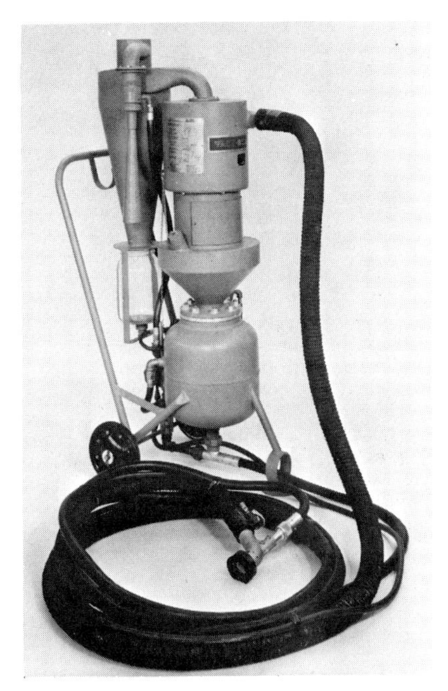

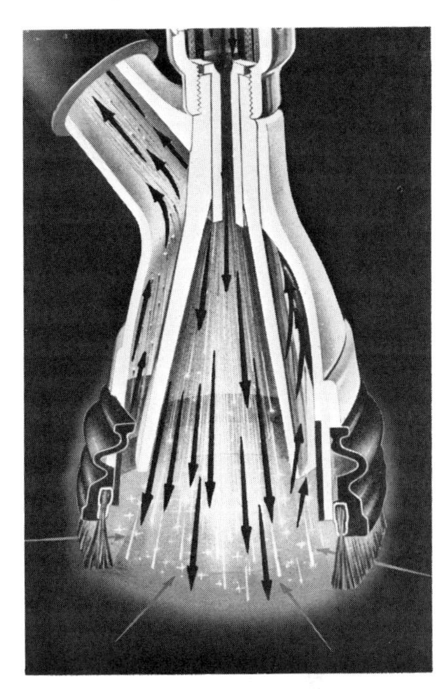

FIG. 4-6 Vacuum sandblast outfit and nozzle. *(Vacu-Blast Corp., Belmont, California.)*

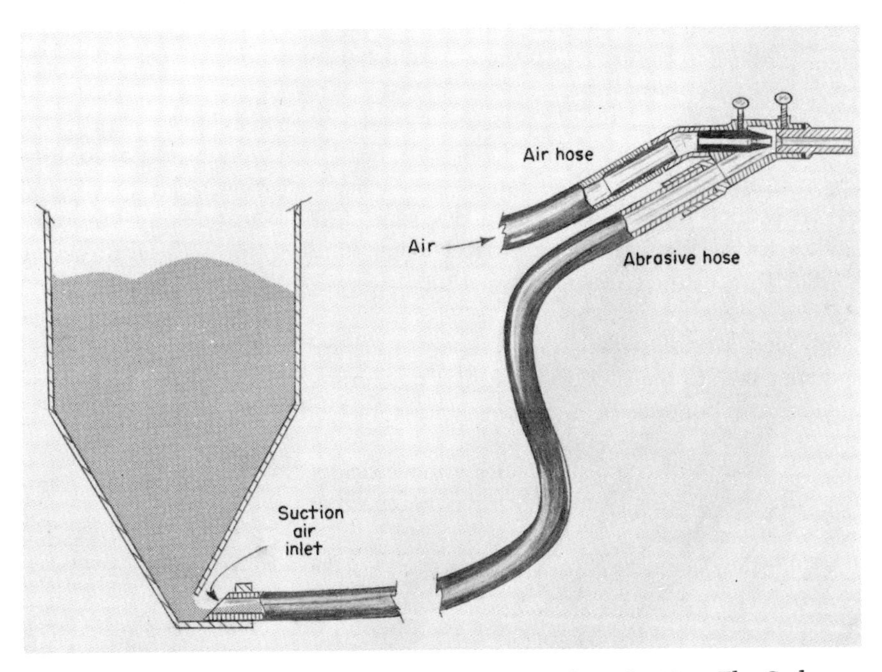

FIG. 4-7 Typical suction-feed sandblast machine. *(Pangborn Division, The Carborundum Co., Hagerstown, Maryland.)*

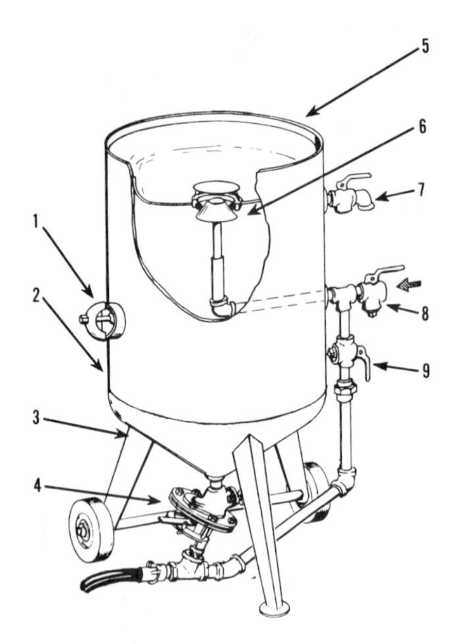

FIG. 4-8 Typical pressure-feed sandblast machine. (1) Handhole; (2) abrasive tank (coded pressure vessel); (3) conical bottom; (4) abrasive flow valve; (5) concave head; (6) automatic pop-up; (7) pressure relief valve; (8) air inlet valve; (9) choke valve. *(Clemco-Clementina, Ltd., San Francisco, California.)*

around machines where dust is objectionable, and for touch-up work.

Suction-feed Sandblaster: Sand is aspirated through the hose from an open hopper. Low cost; simple design; suitable for small jobs or intermittent use.

Pressure-feed Sandblaster: Sand feeds by gravity from a pressure-equalized tank. More efficient; most commonly employed high production blast machine.

To reduce dust hazards from dry sandblasting, water may be added either to the sand or at the nozzle. Wet sandblasting will be discussed briefly toward the close of this section.

The following discussion will be concerned primarily with the pressure-feed sandblast machine, although much of the discussion will apply equally well to the suction-feed blast machine.

The components of a modern pressure-feed sandblast outfit are:

1. Source of compressed air, with air driers
2. Abrasive tank, with abrasive meter valve
3. Blast hose
4. Blast nozzle

Necessary accessories include a pneumatic remote control nozzle (preferably a dead-man control), air-supplied hoods, and protective aprons and gloves.

1. Air Supply: Compressed air supplies the driving force to propel the abrasive particles against the surface with sufficient impact for cleaning. Consequently, having an adequate air supply is one of the most important factors in efficient sandblast operations.

The compressor must have sufficient capacity to maintain a continuous supply of air to the nozzle at 90- to 100-psig nozzle pressure. As an approximation of capacity, a new electrically driven compressor will compress, per horsepower, around 4 cfm free air (air at atmospheric pressure) to a discharge pressure of 100 psig, and 4½ cfm to a discharge pressure of 80 psig. As the compressor parts wear, compression volume will drop to around 75 percent of these values. The capacity of gasoline-driven compressors is approximately one-half of electric compressors with the same horsepower rating.

Table 4-3 presents multiple-stage compressor capacities as related to their horsepower rating, when operated near sea level. Altitude has an appreciable effect on compressor efficiency, and must be considered in determining compressor requirements. Effect of altitude on compressor capacity is presented in Table 4-4.

The air demand of sandblast nozzles is appreciable. An average sand-

FIG. 4-9 Pressure-feed sandblast outfit. *(Sanstorm Manufacturing Co., Fresno, California.)*

TABLE 4-3 Compressor Capacity vs. Horsepower Rating
(Discharge pressure 100–120 psig; continuous operation)

Compressor hp rating	Compressor capacity, free air consumption, cfm	
	Electric drive	Gas engine drive
5	15–20	6–9
7½	20–30	10–15
10	35–45	20–25
15	50–70	25–35
20	75–90	35–45
25	90–105	40–50
30	Up to 130	Up to 60
40	Up to 170	Up to 75
50	Up to 210	Up to 85
60	Up to 260	Up to 100
75	Up to 320	
100	Up to 420	

Note: Larger air compressors of 600, 900, and 1,200 cfm capacity are available, with diesel engine drive.
SOURCE: Air Compressor Research Council; Binks Manufacturing Co., Chicago, Ill.

blasting job on structures or tanks preparatory for painting would call for use of a ⅜-in. nozzle consuming from 170 to 200 cfm free air, discharged at 90 to 100 psig. A 50- or 60-hp (electric) compressor would be recommended.

TABLE 4-4 Effect of Altitude on Compressor Capacity

Altitude, ft	*Loss in capacity*
3,000	10%
5,000	16%
8,000	24%
10,000	30%

For efficient cleaning of steel surfaces, nozzle air pressure must be maintained above 90 psig. At lower pressures efficiency drops rapidly:

Nozzle pressure	*Cleaning efficiency*
100 psig	100%
80 psig	65–70%
60 psig	45–50%

Masonry cleaning, however, requires considerably lower nozzle pressures. Pressures of around 40 to 50 psig would be recommended for most surfaces.

2. *Sand Tank and Flow Valve:* Sand tanks with pressure-feed equipment are pressure-equalized, and so in effect are pressure vessels. It is desirable to have this tank meet the ASME code for unfired pressure vessels. One sand tank will serve for intermittent blasting, while two tanks or a double-chamber tank are required for continuous operation.

Suction-feed sandblast equipment employs a simple open tank with a sand hopper bottom having an air-inlet slot at the sand hose connection. Sand is drawn from the hopper and through the hose by the flow of aspirated air due to the ejection of compressed air through an air jet at the blast nozzle (see Fig. 4-7). The rate of sand feed depends principally upon the size of the air-jet orifice, and this should be slightly less than one-half the diameter of the blast nozzle for proper sand rates.

A critical component of a pressure-feed system is the sand meter/shutoff valve, which controls the sand-feed rate in pressure-feed sandblasters. Proper sand flow control is important in achieving the expected efficiency in sandblasting. Automatic sand meter valves also are available which adjust the sand flow rate as air pressure fluctuates.

It is becoming increasingly necessary to employ a remote control sand valve of the dead-man type, which permits the operator to control sand

FIG. 4-10 Suction-feed sandblast outfit. (*Clemco-Clementina, Ltd., San Francisco, California.*)

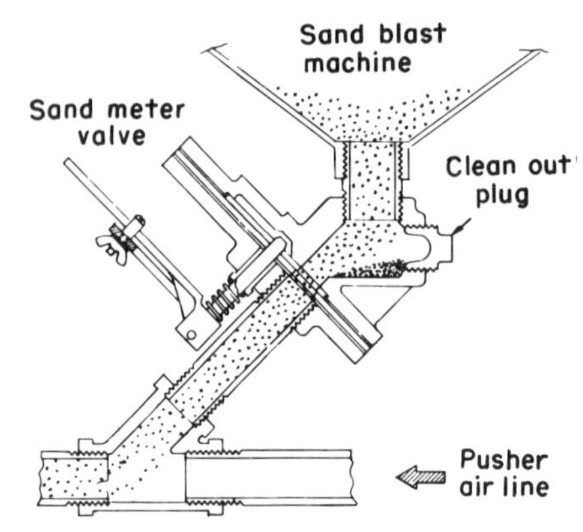

FIG. 4-11 Sand metering valve. *(Clemco-Clementina, Ltd., San Francisco, California.)*

flow at the nozzle. This type of control is now required on all government sandblasting work. It features:

Safety:	Sand flow is controlled at the nozzle, and shuts off if the nozzle is dropped;
Economy:	Saves sand, since sand flow is stopped when blasting stops.

Without a remote control valve, a pot tender must be in constant attendance and alert to the blaster's signals for sand flow changes. Commonly employed signals are shown on page 103.

Visual signals can be made in daylight or when the blaster and pot tender are in sight of each other. Flashing light signals or sound signals usually are used at night. Sound signals can be made with the use of a hammer, horn, or in some cases electric buzzers. They should *not* be made by using the sandblasting nozzle as a hammer.

3. *Hose:* Along with the compressor, the sandblast hose most often is undersized for efficient operation. The air piping and hose should be sufficiently large to minimize air-pressure drops due to friction losses. A 10-psig drop in air pressure means a 15 percent reduction in blast cleaning rate.

Air-pressure losses in hose at different air rates are presented in Table 4-5. These were determined experimentally on pulsating air. Pressure drops at higher airflow rates in larger-size hoses closely approximate the pressure loss in new pipe, as presented in Table 4-6 on page 105.

As a general rule, the hose diameter should be at least three times, but not more than four times, the size of the blast nozzle. If larger than this, the air flow rate may not be sufficient to keep sand from settling. A minimum size of 1-in. ID hose is recommended; 1¼ in. is preferred for the average size job, and 1½- or 2-in. hose should be used for lengths of 100 ft and longer depending upon nozzle size.

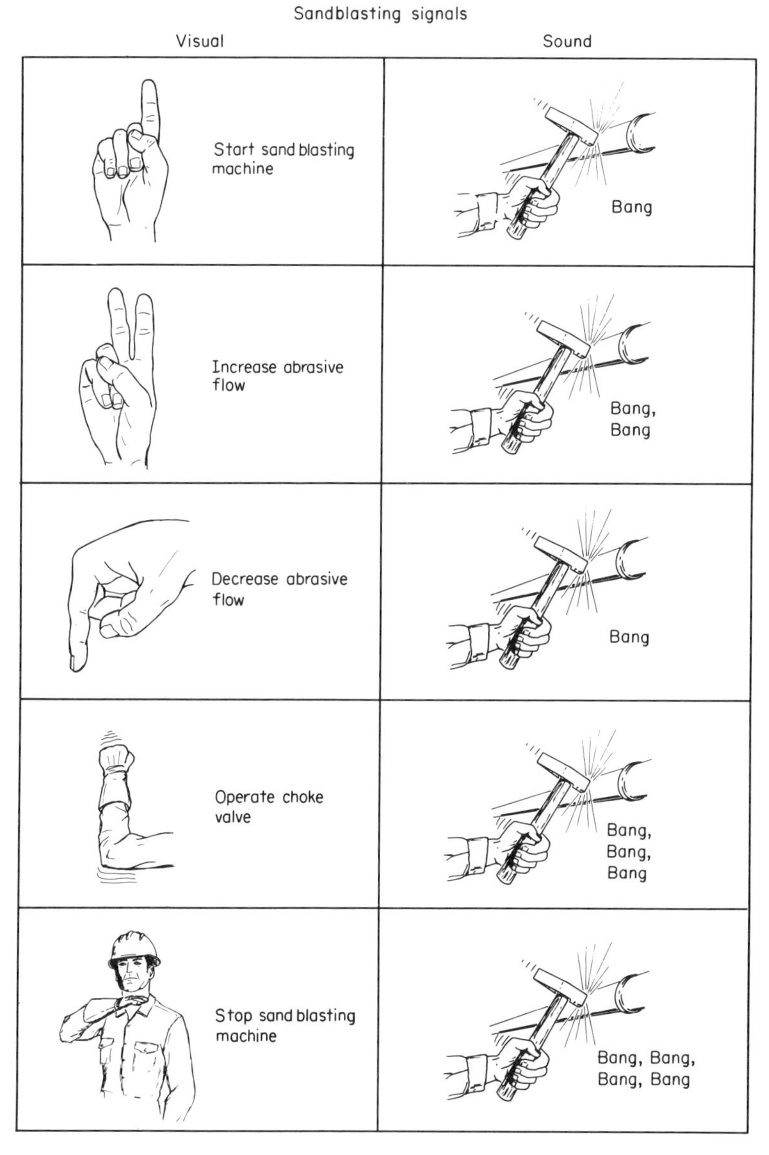

Avoid restrictions in the line by using externally fitted couplings of the 90° turn quick-disconnect type. When whip ends are desired, a two-ply hose of the same diameter may be employed; or preferably, a short length of the next smaller diameter hose is used for increased air velocity to improve sand suspension. In addition, it is convenient to use electrostatically conductive hose to discharge the static electricity generated at the nozzle, thereby eliminating the need for nozzle grounding wires.

TABLE 4-5 Airflow Pressure Drop in Hose

Size of hose, coupled each end	Line pressure, psig	Cfm free air passing through 50-ft lengths of hose					
		40	60	80	100	120	150
		Loss of pressure, psi*					
½ in.	60	8.4	23.4				
	80	6.0	17.4	34.6			
	100	4.8	13.3	27.2			
	110	4.3	12.0	24.6			
¾ in.	60	1.2	2.8	5.2	8.6		
	80	0.8	1.9	3.6	5.8	8.8	
	100	0.6	1.4	2.7	4.4	6.6	11.1
	110	0.5	1.3	2.4	3.9	5.9	9.9
1 in.	60	...	0.6	1.2	2.0	3.3	7.2
	80	...	0.5	0.8	1.4	2.0	3.5
	100	...	0.4	0.6	1.0	1.5	2.4
	110	...	0.3	0.6	0.9	1.3	2.1
1¼ in.	60	...	0.2	0.3	0.6	1.0	
	80	...	0.1	0.2	0.4	0.6	1.0
	100	...	....	0.2	0.3	0.4	0.7
	110	...	....	0.2	0.3	0.4	0.6
1½ in.	60	...	....	0.1	0.2	0.3	0.5
	80	...	....	....	0.1	0.2	0.4
	100	...	....	....	...	0.1	0.2
	110	...	....	....	...	0.1	0.2

* Pressure drops for other lengths are in direct proportion to the change in length: 100-ft lengths have double these pressure drops, etc.

SOURCE: Ingersoll-Rand Company, New York, N.Y.

4. Blast Nozzle: The nozzle is the principal working part of a sandblast machine, and as such is subjected to the greatest wear. Consequently, it has a limited effective life.

The nozzle directs the sand flow into a concentrated stream for most effective cleaning. As the discharge end of the nozzle wears, the sand pattern enlarges and becomes less effective in cleaning. When the discharge diameter has increased 50 percent over the original diameter, the nozzle liner requires replacement.

The inlet end of the nozzle, on the other hand, wears at a very much slower rate. Consequently sand flow and airflow rates will not increase appreciably, and do not serve as reliable indicators of nozzle wear at the discharge end.

In general terms, the tougher the cleaning job, the *longer* the nozzle should be:

3- to 4-in. nozzles for easily removed deposits and scale,

5- to 8-in. nozzles for adherent deposits and mill scale; and
the bigger the scope of the job, where faster cleaning rates are desired, the *larger* the nozzle should be:

¼-in. nozzles for small parts or intermittent use,
⅜-in. nozzles for average production sandblasting,
$\frac{7}{16}$-in. nozzles and ½-in. nozzles for large operations.

Nozzle cleaning rates are directly proportional to their air-handling capacity, all other factors being equal:

¼-in. nozzle:	Cleans area A per hr
$\frac{5}{16}$-in. nozzle:	Cleans 160% A per hr
⅜-in. nozzle:	Cleans 220% A per hr
$\frac{7}{16}$-in. nozzle:	Cleans 320% A per hr
½-in. nozzle:	Cleans 400% A per hr

The venturi-type nozzle is generally accepted as the "standard" production sandblast nozzle. It is appreciably more effective than the cylindrical nozzle in directing a concentrated stream of abrasive particles at a high discharge velocity, and so is 20 to 30 percent more effective in cleaning rates. Sand emerges from a properly operated cylindrical

TABLE 4-6 Airflow Pressure Drop in New Pipe
(May be used to estimate pressure drop in hose)

Pipe diam.	Line pressure, psig	Pressure drop in 50-ft lengths,* cfm free air							
		60	80	100	125	150	200	300	400
¾ in.	80	2.2	3.9	6.0					
	100	1.8	3.2	4.9					
	125	1.5	2.7	4.1					
1 in.	80	0.6	1.1	1.7	2.3	3.8			
	100	0.5	0.9	1.4	2.2	3.1			
	125	0.4	0.7	1.1	2.0	2.6			
1¼ in.	80	0.14	0.25	0.39	0.6	0.9	1.6	3.5	
	100	0.11	0.20	0.32	0.5	0.7	1.3	2.9	
	125	0.09	0.17	0.27	0.4	0.6	1.0	2.5	
1½ in.	80	....	0.11	0.17	0.27	0.39	0.7	1.5	2.9
	100	....	0.09	0.14	0.22	0.32	0.5	1.3	2.3
	125	....	0.07	0.12	0.18	0.26	0.5	1.0	1.9
2 in.	80	....	....	0.05	0.07	0.10	0.19	0.4	0.7
	100	....	....	0.03	0.06	0.09	0.15	0.34	0.6
	125	....	....	0.03	0.05	0.07	0.12	0.28	0.5

Reprinted from the "Compressed Air and Gas Handbook," 3d ed., Copyright 1961 by the Compressed Air and Gas Institute, 122 East 42nd St., New York, N.Y. 10017.

* Pressure drops for other lengths are in direct proportion to the change in length: 100-ft lengths have double these pressure drops, etc.

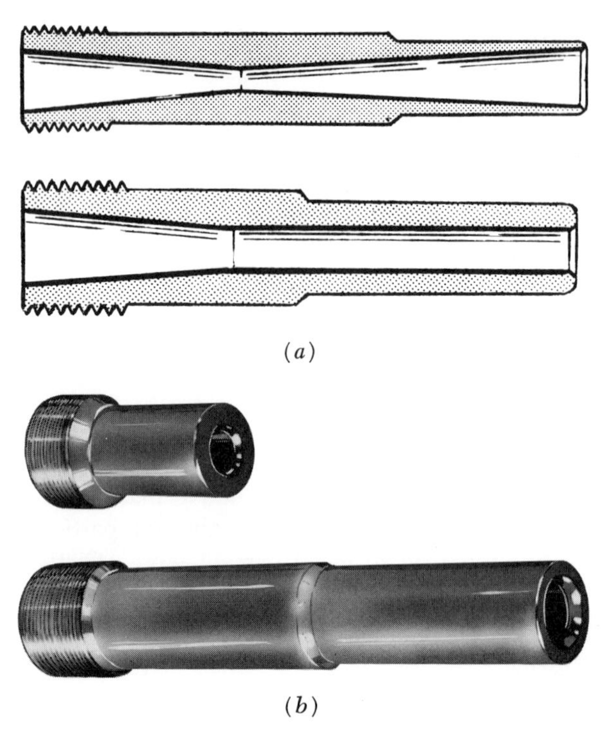

(*a*)

(*b*)

FIG. 4-12 Typical sandblast nozzles. (*a*) **Venturi and conventional nozzles;** (*b*) **short and long nozzles (threaded type).**

nozzle at 210 to 220 mph, and from a venturi nozzle at around 450 mph. At these higher speeds sand shape is not important; round beach silica sand will be as effective an abrasive as inland fractured sand grains.

Nozzle liner wear rates depend upon both the abrasive and the type of liner material. Lighter abrasives such as sand cause greater nozzle liner wear than do denser materials, and finer mesh size abrasives cause more rapid wear than do coarser sizes. As for liner construction, only tungsten carbide and boron carbide are considered for production sandblasting equipment. As the following table shows, other materials have too short a life to be economical except for small-scale intermittent work.

Sand- and air-consumption rates depend primarily on the size of the blast nozzle. Doubling the size (diameter) of the nozzle will increase by four times the rate of air and sand consumption. Air pressure at the nozzle also has a major effect on both air and sand rates. However, although lower rates are shown for lower air pressures, actual amounts of sand and air consumed will be greater because of lowered cleaning efficiency.

TABLE 4-7 Air-blast Nozzle Liner Life

Liner material	Useful life, hr	
	Sand	Steel
Ceramic	1–2	3–5
Cast iron; manganese steel	6–8	15–20
Tungsten carbide	250–300	700–800
Boron carbide	750–1000	2000+

SOURCE: Clemco-Clementina Ltd., San Francisco, Calif.

Approximate sand and air rates and corresponding compressor requirements are presented in Table 4-8. The values given for sand consumption in this table may be used to estimate rates for other abrasive materials by applying the following multipliers:

Crushed flint: ×0.9
Sand, silica: ×1.0
Crushed slag: ×1.1
Garnet sand: ×1.4
Aluminum oxide: ×1.5
Steel grit: ×2.5

Equipment operation and safety: Sandblasting is the quickest, most convenient, and most thorough method of surface preparation available to the painting contractor. The equipment is portable; it is equally suitable for shop cleaning as well as for field use; sand is low cost and usually is locally available; and the abrasive blast can be pinpointed as desired.

There are, however, certain very important precautions and limitations in sandblast operations, and these are reviewed below:

The air employed in sandblasting must be dry and oil-free. Oil and water separators must be well designed, and attended as necessary. This item of equipment usually is neglected, with the very undesirable consequence that condensed moisture and compressor oil are blown upon the surface being cleaned. This probably will be first noticed when the air stream is used to dust off the surface preparatory to painting.

In sandblasting, the nozzle position is varied depending upon surface conditions. The usual rust scales and deposits are removed at a nozzle-to-surface distance of 18 inches, while tightly adhering mill scale and rust require a closer nozzle. Most deposits are effectively removed at a 60 to 70° nozzle angle; pitted surfaces are cleaned at an 80 to 90°

TABLE 4-8 Blast Nozzle Air and Sand Rates

Nozzle		Nozzle air pressure, psig			
		70	80	90	100
⅛ in.	Air rate*	15	17	18½	20
	Sand rate*	110	125	140	160
	Comp. hp*	5	5	7½	7½
3/16 in.	Air rate	33	38	41	45
	Sand rate	250	270	300	350
	Comp. hp	10	10	15	15
¼ in.	Air rate	61	68	74	81
	Sand rate	450	525	575	650
	Comp. hp	20	20	25	25
5/16 in.	Air rate	101	113	126	137
	Sand rate	775	850	950	1050
	Comp. hp	40	40	50	50
⅜ in.	Air rate	143	161	173	196
	Sand rate	1100	1200	1325	1450
	Comp. hp	60	60	75	75
7/16 in.	Air rate	194	217	240	254
	Sand rate	1500	1650	1800	2000
	Comp. hp	75	75	100	100
½ in.	Air rate	250	280	310	340
	Sand rate	1900	2100	2300	2500
	Comp. hp	100	100	150	150

* Sand rate: lb/hr. Air rate: cfm free air. Air rate is adjusted for sand presence. Electric hp requirement is given for recommended compressor size. (All values are approximate.)

SOURCES: Clemco-Clementina, Ltd., San Francisco, Calif.; Sanstorm Manufacturing Co., Fresno, Calif.

angle, whereas old paint films can be lifted best at a 45 to 60° blast angle.

It is important when using hard, dense abrasives or coarse sand to be careful that the surface is not gouged too deeply. Formation of too deep anchor patterns on the surface is very undesirable from the standpoint of coating life, and once formed there is no easy way to reduce their depth.

Avoid all contamination and wetting of the blast-cleaned surface. Finger prints are especially detrimental because of the oils and corrosive salts left on the surface.

The blast-cleaned surface should be treated or primed as soon as possible, and in any case before any visible rusting occurs. The permissible time lag between blasting and priming of the surface will depend upon the steel temperature and relative humidity of the air. Under normal conditions priming should follow no later than 8 hours after sandblasting, and never should be left for the next day.

Sandblasting is potentially a very hazardous operation. The sandblast particles emerge from the nozzle at high speeds and, if directed toward a person, can easily maim or kill. A dead-man type of nozzle control, which automatically stops abrasive flow if the nozzle is dropped, is essential for safe sandblasting.

Dust generation is probably the most serious problem in sandblasting. The operators must be protected against silicosis and other less serious dust hazards through use of well-made air-supplied hoods and heavy-duty aprons and gloves. Nearby workers should wear dust filter respirators. Flying dust and sand prohibit any nearby painting, and nearby machinery must be well protected against sand in moving parts.

Where presence of a flammable mixture may exist, and also for psychological reasons, the blast nozzle should be grounded. In some cases static sparks may be caused by ungrounded sandblast nozzles. These

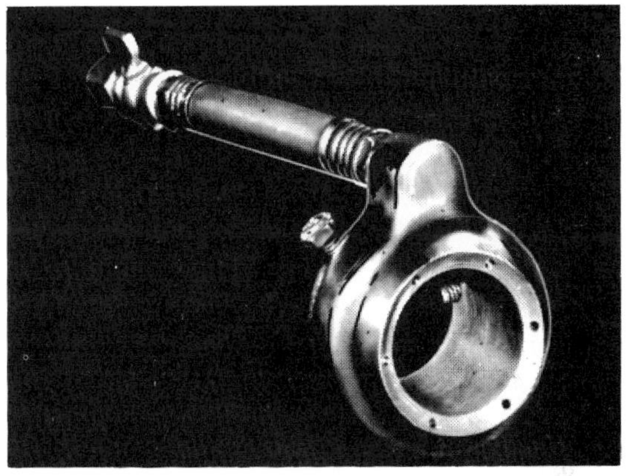

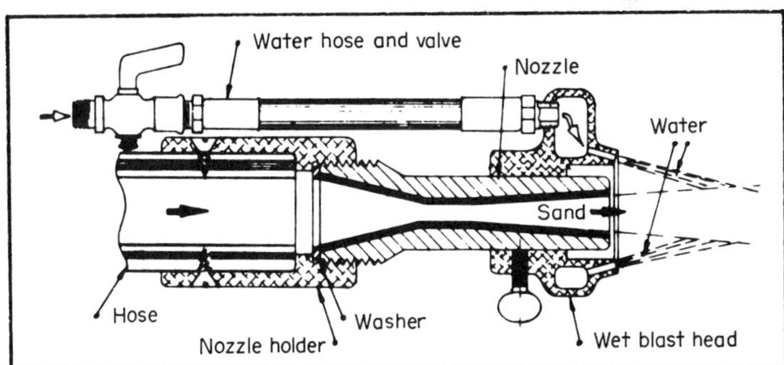

FIG. 4-13 Typical wet blast nozzle attachments, employed to minimize dust from sandblasting. (*a—Clemco-Clementina, Ltd., San Francisco, California.*)

are dangerous as sources of ignition but only uncomfortable or disconcerting to personnel. They do not constitute a safety or health hazard as such.

Wet blast cleaning: Water sometimes is added to the abrasive stream to dampen or wet the sand for the purpose of reducing or eliminating dust formation. This is common practice when sandblasting masonry or cement surfaces. It is less often applied to steel surface preparation due to rusting because of the water. Rust inhibitors must be added to the water to protect the metal surface while wet.

Inhibited water may be mixed with the sand in the sand tank of pressure-feed equipment using approximately 15 gal water to 1 ton sand, but untreated water is more commonly injected at the blast nozzle. Special nozzles, as well as wet blast head attachments for standard dry blast nozzles, are available to inject the water at either the nozzle inlet or discharge end. Injection at the nozzle permits control of water volume to provide just the degree of dampness required to settle dust without excessive wetting of the surface.

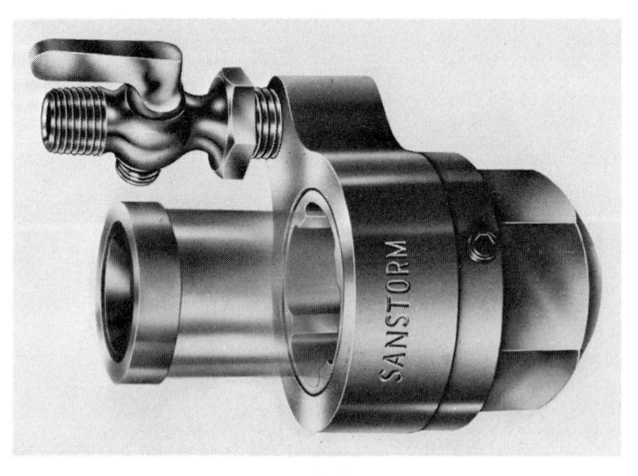

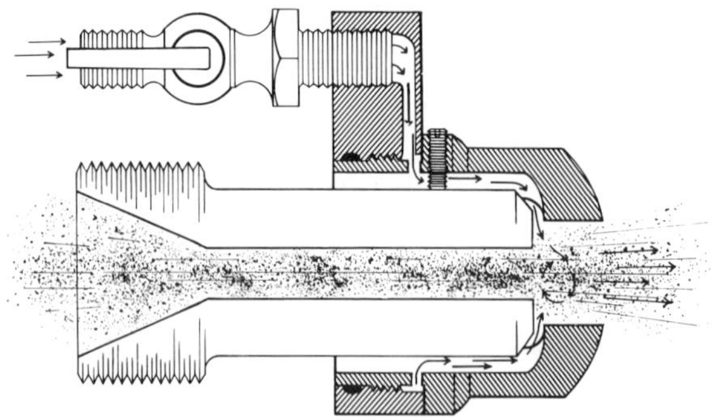

FIG. 4-13 (Continued) *(b—Sanstorm Manufacturing Co., Fresno, California.)*

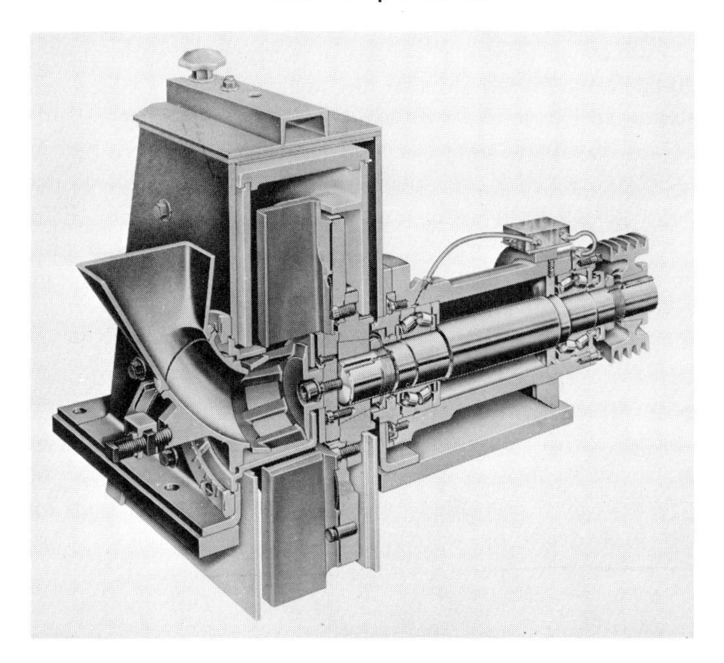

FIG. 4-14 Wheel blast equipment and blast wheel. (*Pangborn Division, The Carborundum Co., Hagerstown, Maryland.*)

When wet-blasting a vertical surface, considerable water will run down the side and slop on the surface below. When wet-blasting horizontal surfaces, the wet sand will pile up into a solid mass directly before the blast operator and must be removed by shoveling.

Soon after the wet blast operation is completed, the surface must be hosed off with water containing a rust inhibitor. Effective inhibitors are 0.2 percent by weight solutions of chromic acid or sodium dichromate or a 2 percent solution of a mix of 4 parts diammonium phosphate with 1 part sodium nitrite.

Wheel blast cleaning: With wheel blast cleaning, the abrasive particles are propelled against the surface at high velocity by means of electrically driven wheels. No compressed air is employed. The wheels are positioned in a cabinet so that all surfaces are cleaned while the article moves on a conveyor or rotary table.

Wheel blast equipment is used only in plant or shop for surface-cleaning and descaling operations. The equipment is compact, self-contained, easy to operate, and provides any desired grade of surface preparation at economical operating costs. Horsepower requirements are about 10 percent of that required for air blast equipment of the same capacity.

Wheel blasting is essentially a dust-free operation. Metal shot and grit are commonly employed, which are reclaimed, air cleaned, and returned to the bin for reuse.

Chemical and Solvent Cleaning

Introduction: Chemical surface-cleaning methods grouped in this discussion include: solvent degreasing; washing with soaps, detergents, alkaline cleaners, and emulsifiable solvents; steam cleaning; acid wash; proprietary rust removers; and chemical paint removers. Each of these methods was designed for removal of specific types of surface contaminants under limited operating conditions. The method used should be chosen to suit the type and amount of surface contamination to be removed.

Chemical surface-cleaning methods are widely used in field operations to remove oil, grease, dirt, old paint, light rust, and sometimes to loosen heavy scale. Solvent degreasing often is required as a preliminary to subsequent mechanical cleaning.

These methods all have one serious shortcoming when employed alone for surface preparation: they do not produce the anchor pattern so desirable for paint application.

Removal of mill scale by pickling is a general shop-cleaning method.

This is the preferred method of surface preparation for galvanizing. Pickling followed by a phosphatizing pretreatment also provides a superior surface for protective coatings.

Solvent degreasing: Degreasing operations are almost always specified in conjunction with some other more thorough surface-preparation method. Degreasing may be accomplished by:

Wiping: Solvents that may be used for wiping include stoddard solvent, petroleum naphtha, mineral spirits, turpentine, xylene, and toluene. The last two solvents are used when greater solvency is required, but they are more toxic. Benzene (benzol), gasoline, and chlorinated hydrocarbons are dangerous and never should be used for wiping.

Cloth rags, sponges, and brushes are dipped in the solvent cleaner and rubbed against the oily surface. Excess solvent must drain from the surface for removal of contaminations. It is important that the last wash or rinse be made with fresh solvent and clean rags. Vertical or sloping surfaces shall be cleaned from the top downwards.

Spraying: Solvent cleaner may be spray-applied under pressure. Excess solvent is collected into a reservoir or sump in order to be reused. Final spray must be made with clean solvent.

Vapor Degreasing: Special vapor degreasing equipment permits trichlorethylene or perchlorethylene to be heated to the boiling point, thus generating a vapor zone. The parts are suspended in this vapor zone, where they are washed with pure solvent that condenses on all surfaces.

Paint stripping; alkali cleaning: Nonflammable, solvent-type paint removers are based on strong chlorinated solvents containing wetting agents, and sometimes a minor amount of paraffin wax to reduce loss by evaporation. The remover is applied by brush, and the softened paint must be scraped and wiped off. It is essential that all residue paint remover be washed off the surface with an aromatic hydrocarbon solvent (xylene, toluene, or enamel thinner), otherwise the subsequently applied paint coating will not dry and adhere properly.

Alkaline-type Paint Strippers: Most paint, lacquer, and varnish coatings are easily removed by hot solutions of caustic soda or trisodium phosphate to which starch, bolted whiting, or sawdust is added to thicken the solution. This mixture is applied hot by using a long-handled brush, by spray, or by flooding. The softened paint is then scraped or rubbed off, or blown off with steam. The caustic solution should be completely removed by repeated rinsing with acidulated water, giving particular attention to all cracks and crevices.

Very effective proprietary alkaline paint-stripping and surface-cleaning formulations are available. These contain blends of strong alkalies, detergents, and wetting, emulsifying, and saponifying agents which remove old paints, oils, grease, dust, dirt, and much of the loose rust.

Dilute solutions of these cleaners are applied hot (nearly boiling) to vertical surfaces by means of perforated pipe or rakes. About 1½ gal per min is applied per foot span of surface and allowed to cascade downward. For most economical operations, the solution is collected at the bottom and returned to the heating tank for reuse.

Upon completion of the cleaning operation, the surface must be thoroughly rinsed to remove all traces of the alkali solution, followed by an acid-water rinse to prepare the surface for painting. Dilute solutions of chromic or phosphoric acids are suitable for this final acid rinse.

Steam cleaning: Steam cleaning augmented with the detergent action of cleaning compounds is a very effective method for removing oils, grease, and dirt contaminations. This method is widely employed

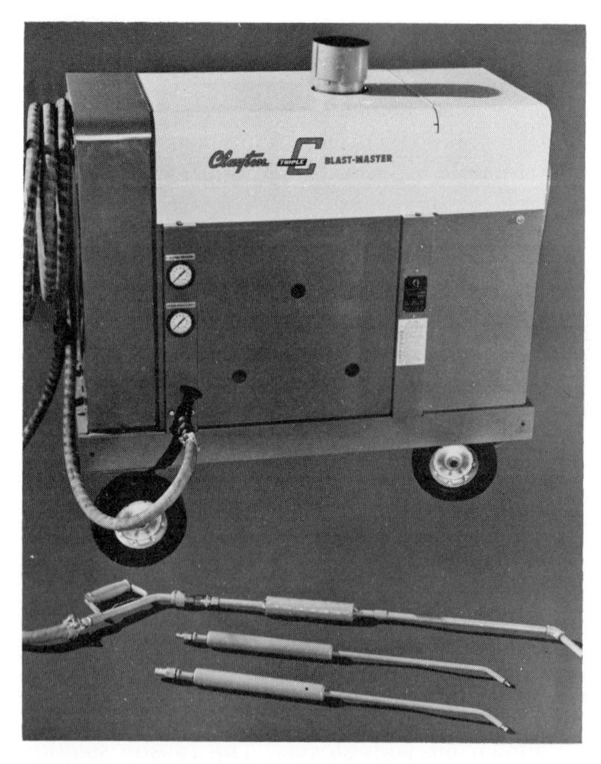

FIG. 4-15 Self-contained steam cleaning outfit. (*Clayton Manufacturing Co., El Monte, California.*)

for surface cleaning of machinery, mobile equipment, and other surfaces contaminated principally by deposits of grime, dirt, and those of an oily nature.

Professionally designed self-contained steam cleaning outfits are available which consist of an oil or gas-fired steam-generating unit with suitable controls and gauges, steam hose, and a gun with interchangeable nozzles of various shapes.

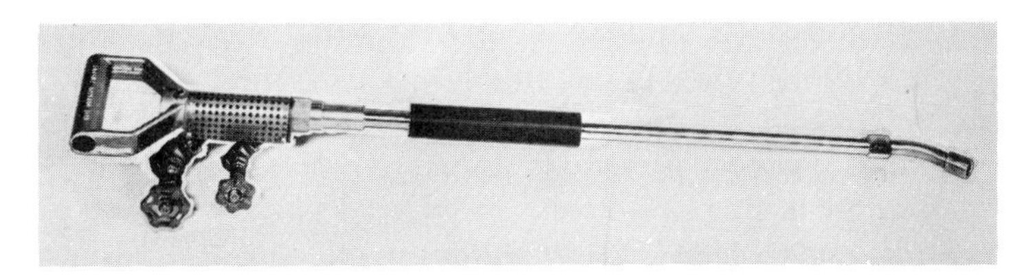

FIG. 4-16 Modern two-feed steam cleaning gun. (*Oakite Products, Inc., Berkeley Heights, New Jersey.*)

In some models, the volatile cleaning compound is added to the steam-generating water, while others employ guns with two separate feeds—for steam and for hot water with detergents. Controls allow the operator to apply each feed separately or together. Still another model permits a quick change in the feed to the gun from steam with detergent, to high pressure hot water with or without cleaning agents. For detailed information on steam cleaning equipment, the reader should refer to the manufacturers' literature.

Steam or hot water is applied under pressure against the work using a gun fitted with an appropriate nozzle. For broad surfaces the fan nozzle is used; for smaller inaccessible areas the round nozzle is better. Nozzles with a fiber brush attached are employed if some additional scrubbing is desired.

The surface should first be wetted to allow the cleaning compound to loosen the soiling material. This loosened material is then removed by the scrubbing action of a steam pass, which also forces the cleaning solution ahead of the steam contact area.

The speed of cleaning depends upon the degree of surface contamination. The cleaning should be carried out systematically from top to bottom of the work. A final rinse with steam or hot water is required, using acidulated water if strong caustics were employed for cleaning.

Following hot-alkaline or steam cleaning, a preferred pretreatment for painting would be the application of a hot phosphatizing solution by means of the two-feed steam gun. The solution is applied on vertical surfaces from the bottom upwards, followed by a water rinse and steam dry.

Acid cleaning: Acid cleaning solutions are most often applied to prepare aluminum or galvanized surfaces for painting. These usually are based on phosphoric or dilute hydrochloric acid solutions, and produce a clean, finely etched surface suitable for paint application.

Acid cleaners are occasionally applied to steel surfaces. Concentrated metal cleaners, containing approximately 60 percent by volume of phosphoric acid (85 percent), are diluted with approximately three times their volume of water. Acid-resisting containers should be used. The diluted cleaner is applied liberally to the work with a long-handled brush while scrubbing thoroughly. This is repeated if necessary until all rust and oil are removed. The work must then be thoroughly rinsed with clean water, preferably by hose. If possible, hot water should be used for rinsing, which will speed up drying.

Safey and Health Considerations: All chemical surface-preparation methods described above involve more or less severe health hazards and/or fire hazards. Overspray and runoff of cleaning solution in case of overhead cleaning operations constitute a special safety hazard. Warning signs should be placed and equipment below should be adequately protected.

Eye and skin protection are always required when working with caustics, acids, steam, or solvents. The use of volatile solvents calls for adequate ventilation and other fire-prevention measures. Refer to Chap. 3 of this manual, where safety considerations are covered in greater detail.

Metal Pretreatment Processes

Pretreatment of a cleaned metal surface sometimes is specified or recommended to improve further the condition of the surface for paint application. Pretreatments usually provide some rust protection to the freshly prepared surface, as well as improve the adhesion of the primer.

Pretreatment with a wetting oil very often is specified for oleoresinous paints to be applied to surfaces prepared by less thorough methods than abrasive blast cleaning. Application of a wash primer to sandblasted surfaces usually is recommended to improve adhesion of vinyl coatings. Phosphatizing is an excellent shop pretreatment method for most types of protective coatings. These three principal pretreatment processes are discussed in this section.

Wetting oil pretreatment: Application of wetting oils often is specified for hand- or power-tool-cleaned surfaces and sometimes with brush-blast surfaces, where appreciable amounts of adherent rust scale, mill scale, and old paint remain. Wetting oils are used only in conjunction with oleoresinous primers, and are especially recommended before application of medium and short oil phenolic paints. They cannot be used under the synthetic resin lacquer-type paints such as vinyls or chlorinated rubber, or under heavy-duty protective coatings.

Wetting oil is prepared by thinning 1 part of raw or boiled linseed oil with from 1 to 2 parts of turpentine or mineral spirits. Boiled linseed oil is used when the drying time of the primer is to be held to a minimum. Some proprietary wetting oils based on processed fish oils also are available and give excellent results.

The wetting oil shall be applied freely by brush or spray so as to saturate the surface. Dull spots indicate that all oil has been absorbed and these should receive an additional thin coat. Excess oil should be wiped from the surface before it dries.

Any rust or scale which has been loosened by the wetting oil shall be removed by scraping or wire brushing. Dry spots so exposed shall again be treated with wetting oil.

The primer may be applied as soon as the wetting oil has sufficiently set to permit normal application, but should not be delayed more than 3 or 4 hr after application of the wetting oil.

Wash primers: Vinyl butyral wash primers originally were developed for application to sandblasted steel to assure adhesion of vinyl coatings. However, they have been found to aid in the adhesion of most other paints as well, including cold-applied asphalt and coal-tar coatings.

Wash-primer pretreatment also is recommended for galvanized steel, aluminum, tin, terneplate, and copper in preparation for painting. In addition, wash primers serve as an excellent temporary protective primer for blast-cleaned surfaces which for construction reasons cannot be painted immediately.

The wash primer is prepared soon before its application by mixing the two components, pigment base and acid diluent, which are supplied in separate containers. The pigment base consists essentially of a solution of polyvinyl butyral resin in alcohols, pigmented with basic zinc chromate. The acid diluent contains phosphoric acid, water, and alcohol. Four parts of the pigment base are mixed with one part acid diluent. Wash primer that cannot be used within a maximum of 6 hr after mixing must be discarded.

The wash primer is applied by spray or brush at a spreading rate of about 250 to 300 sq ft per gal. Spraying is generally preferred. Avoid excess air pressure or liquid pressure. Apply a wet coat; if dusting occurs, move the gun closer to the surface or decrease the atomizing air pressure. No thinning is required.

The wash primer will dry in ½ to 1 hr. The dry film thickness should be only 0.3 to 0.5 mils. The normal appearance is an uneven coloring, which does not completely hide the base metal. Use denatured or isopropyl alcohol to clean equipment and tools or to remove deposits of excess wash primer.

This pretreatment is not to be considered as a coat of paint. Painting should follow as soon as practical, but may be delayed several days.

Phosphate coatings: Application of phosphatizing solutions is recommended for chemical, solvent, and steam-cleaned surfaces, as well as for all surfaces which will be subjected to severe service exposures. Phosphatizing is primarily a shop pretreatment method; however, the

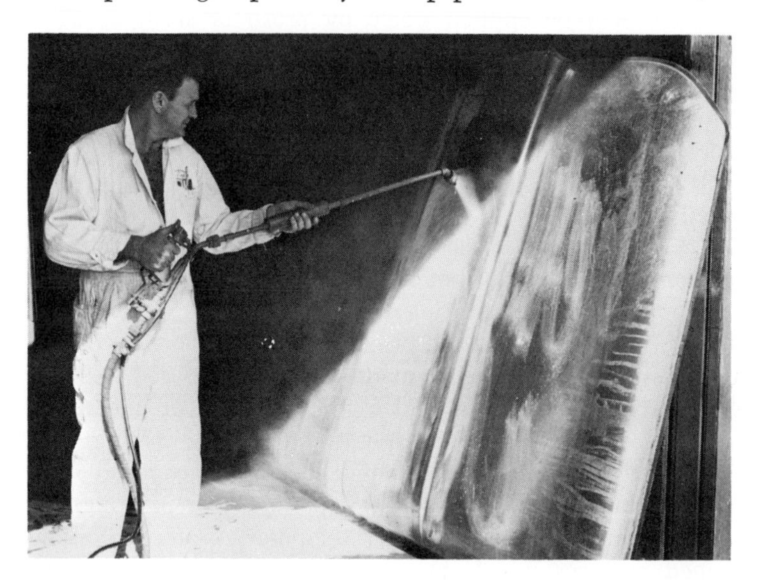

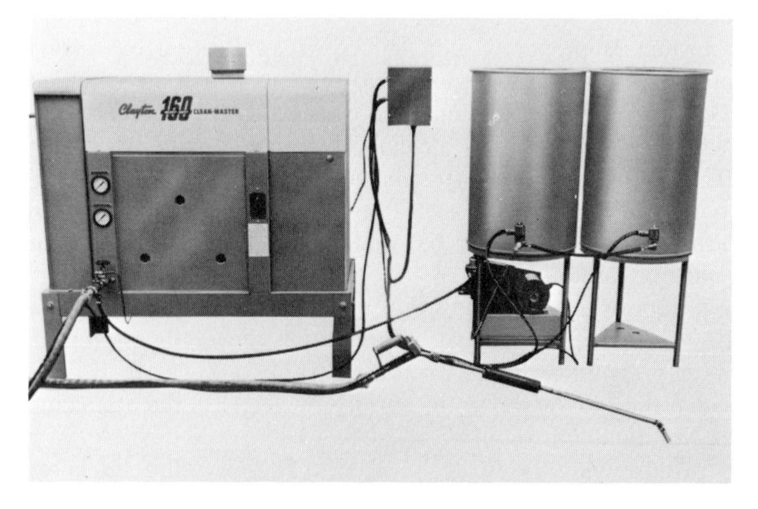

FIG. 4-17 Steam cleaning and phosphating equipment. (*Clayton Manufacturing Co., El Monte, California.*)

more versatile models of steam cleaning equipment make possible the easy application of hot phosphatizing solutions to field structures, tanks, etc.

Phosphatized surfaces extend paint service life because:

1. Paint adhesion is improved, which is important where the surface-preparation method did not provide an anchor pattern.
2. Undercutting corrosion is retarded by the corrosion-preventive phosphate film.
3. The paint is electrically insulated from the metal surface, so galvanically accelerated corrosion does not occur at breaks in the paint film.

Two types of phosphate coatings are suitable for field application. One consists of an iron phosphate coating which is formed on the steel surface through application of a hot 10 to 12 percent solution of phosphoric acid containing emulsifying and wetting agents, detergents, etc. The solution is allowed to dry on the surface, after which it is rinsed off preparatory to painting.

Another process produces an iron-zinc phosphate coating on the steel surface through the application of a solution of phosphoric acid saturated with acid-zinc phosphate, together with the usual wetting agents and reaction-modifying chemicals.

Excellent proprietary formulations are available of both types of phosphate pretreatment coatings discussed.

In addition to the pretreatment of steel surfaces, application of zinc phosphate coatings has been proven to be the most effective method for preparing zinc and galvanized surfaces for painting. Special proprietary formulations also are available for this purpose.

Preparation of Nonmetallic Surfaces

Surface preparation of wood, plywood, wood fiber board, gypsum wallboard, plaster, concrete, and masonry for painting or repainting involves the removal of all loose, flaky, or otherwise unsound material and of oily or greasy contaminants which adversely affect adhesion. Included in this discussion are methods to neutralize, etch, or roughen nonmetallic surfaces.

The method or methods to be used depend largely on the type and condition of the surface to be painted and on the type of paint selected. The methods described below will serve as guides. Experience and skills developed by the painter should determine the proper choice of preparation method to use in each individual case.

Wood, plywood, and wallboard: Wood that is to be painted should be well seasoned and dried to a moisture content of around 15 percent.

If the moisture content is much in excess of this figure, the paint tends to blister or peel. The moisture content may be measured by an electrical moisture meter.

Remove loose knots and fill knot holes with plastic wood, or cut out and fill with sound wood. Any pitch exudations on the surface should be removed by scraping and sanding, or if the resin is soft by cleaning with mineral spirits or turpentine. Any loose grain or splinters should be removed.

The surface should be planed and sanded as required by the need for good appearance. Interior woodwork usually requires a smoother surface than most exterior woodwork. Sandpaper obliquely across the grain for painting, and with the grain for staining or natural finishes. Set nails well below the surface to provide a firm key for the filler. Remove contaminations of oil, asphalt, dirt, or crayon marks on lumber by scraping, sandpapering, or use of solvents.

Knots, resinous streaks, and discolored sapwood should be sealed with two thin coats of shellac extending about an inch beyond the actual area of the defects. If the lumber has been treated with a preservative that has been found to affect the adhesion of paint, apply a full coat of wash primer over the entire surface.

Fill nail holes, cracks, joints, and other irregularities of the surface after priming. When done before priming, the filler may shrink and fall away due to absorption of some of the binder by the wood. The filler shall be rubbed down to a level surface with sandpaper or pumice stone.

Soft insulating boards should be dusted prior to painting. Because water-base paints normally will be used, nails should be galvanized or blued. Hardboard surfaces should be dusted and contaminations removed by sanding or by solvent. Fill nail holes after priming.

Gypsum wallboard surfaces usually have taped joints. Set the tape in joint cement forcing the cement through holes in the tape, and fill the recess neatly with joint cement. Feather edges. Also fill intermediate nail holes. Sand joint and nail holes smooth when dry. Dust the surface just prior to painting.

Plaster and concrete surfaces: The principal requirement for successful painting of plasterwork is the proper curing and drying of the plaster. Painting should be delayed as long as possible. Under favorable drying conditions, 30 days are considered a minimum. Plaster on new brickwork or building block takes longer to dry thoroughly than plaster on metal lath. It should be remembered that a plastered wall seldom dries uniformly.

If any salt deposits (efflorescence) are visible, the areas should be scrubbed with a stiff dry brush and rubbed with a damp (not wet) cloth to remove the salts, and should then be left for a week. If necessary, this must be repeated until no more salts appear. As long as efflorescence develops, it is an indication that water is moving towards the surface.

If a surface persistently remains damp, which may be due to hygroscopic salts, the cause should be sought and corrected before attempting to apply paint.

Lime plasters, portland cement plaster, and gypsum plasters containing lime are alkaline. Unless thoroughly dry and matured, they will cause an alkali deterioration of oil-base paints. A small patch test will help to determine when the risk of alkali attack has reached a minimum.

New concrete and masonry should be allowed to cure and dry for several months before painting. No paint, except water cement paints and certain water-base paints, can be applied safely to a damp surface. Surface alkalinity and movement of moisture through the body of the concrete attack and deteriorate oil-containing paints.

When prolonged aging is impractical, the surface alkalinity may be reduced by treating the surface with a water solution containing 3 percent phosphoric acid and 2 percent zinc chloride. The concrete should be at least three weeks old. After drying not less than 24 hr after treatment, all loose particles must be brushed off. Alkali-resistant paints may then be applied when the surface is dry, although the body of the concrete may still contain a large amount of moisture.

An effective seal coat for plaster and concrete surfaces is one or two coats of a water-base PVA or acrylic paint. These often are applied to damp surfaces, since they are unaffected by alkalies. It is common practice to apply such a coat before finishing with an oil-base paint.

Surface contaminations such as oil from the concrete forms, grease, or salt deposits must be removed by solvents, steel brushes, abrasive stones, or preferably by light sandblasting. The surface shall be rough enough to establish a good mechanical bond with the coating. This may be accomplished when necessary by etching with diluted (15 to 20 percent) hydrochloric acid followed by thorough flushing with water.

Old concrete and masonry should be dusted, scrubbed, wire brushed, or sandblasted as required to remove surface contaminants and to provide a satisfactory key for paint.

Previously painted surfaces: Old paint work that is in a sufficiently sound condition not to require removal should be cleaned of accumulated dirt by solvent cleaning, washing, or simply by dusting, and shall be rubbed down to abrade the surface for proper adhesion of the new paint. For this purpose pumice stone or waterproof sandpaper is used. Any mold should be killed by applying an antiseptic wash.

Old coatings which have failed extensively or show signs of general poor adhesion must be completely removed either by burning or by use of caustic- or solvent-type paint removers.

When applying a high quality paint over existing paints of unknown composition, a small patch application first should be made to see if the solvents will wrinkle or lift the old paint. Heavy-duty paints with strong solvents cannot be applied over old oleoresinous paints for this reason.

Chapter 5

Paint Application Methods

General Considerations in Paint Application

Experience has proven that in order to obtain the desired paint life or coating service, proper application is just as important as are satisfactory surface preparation and the correct choice of paint. The experienced painter must be familiar with the different methods of paint application and recognize the advantages and limitations of each.

This chapter may be used as an outline for training painters in these principal methods of paint application: spraying by air atomization, airless, and electrostatic methods; brushing and roller coating; and daubing and troweling of heavy-bodied materials. In addition, this chapter will serve as an aid to construction engineers and other supervisory personnel in scheduling paint work as an integral part of construction.

Although paints are applied by spraying in almost all industrial or large-scale applications, spray painting is becoming increasingly popular for smaller jobs as well. Brushing is employed by professional painters only where spraying is unsuitable: for interior painting of residences, trim work, etc., and where paint fog is objectionable. Roller coaters are commonly used for applying water-base paints to interior walls. The amateur do-it-yourself painter almost always paints by brush or roller coater, principally because of the low equipment cost.

Paint wastage through spray painting is high, but because of the speed of application paint loss is more than compensated for by savings in labor. Electrostatic spray application, however, combines speed with low paint loss, and so there is increasing interest in its use for both in-plant and exterior paint application.

Mixing and thinning: All paints will show some separation of the lighter and heavier components during storage. Some pigments settle to a hard cake and some to a soft sludge. This property determines the shelf life of the paint.

The shelf life of some paints is further limited by a tendency of certain binders to curdle or gel upon prolonged storage. This deterioration should not be confused with the highly thixotropic properties of gelled oil binders, which become homogeneous and free-flowing upon stirring.

TABLE 5-1 Estimates of Paint Loss During Application

Application method	*Paint loss*
Conventional air spraying	20–40%
Hot air spraying	15–30%
Airless spraying	10–20%
Electrostatic spraying	5–15%
Brushing and roller coating	4–8%

If the paint cannot be broken down by stirring to a homogeneous liquid state, it must be discarded.

Many paints form a skin over the liquid surface while in the can. The skin should be carefully cut loose from the side of the can. A firm skin may then be easily lifted out in one piece. If the skin is weak or brittle, first remove as much as possible and then strain the paint through a fine screen or cheese cloth after stirring.

This skin represents loss of binder. In the case of oleoresinous paints, addition of a small quantity of boiled linseed oil will compensate for this loss.

Hand mixing may be employed for containers less than 5 gals in size. The best method is to pour off the thin portion into a clean container, then mix the settled portion with a strong paddle, breaking up any lumps by rubbing them against the side of the can. When the settled portion has been worked into a smooth paste, gradually return the decanted portion with continuous stirring. Finally, box the paint by pouring several times back and forth from one can into the other. The same method is used for adding aluminum paste or other pigment pastes to varnishes or paints.

Mechanical mixing should always be employed for 5-gal-size containers or larger, and also for mixing relatively small amounts of catalyst into viscous paints or mastics. The mixing should be by propeller, screw, or paddle-type agitator, driven by compressed air or electric motor.

Mixing of unopened cans of paint in 1-gal or smaller sizes may also be performed by mechanical mixers which rotate, shake, or vibrate the container. The time necessary to mix the paint thoroughly must be determined by experience. This method should not be used for emulsion paints or clear varnishes, because objectionable air bubbles will be introduced into the material. These bubbles escape very slowly.

Thinning should be supervised by a competent person. Only the minimum amount of the specified thinner should be added to obtain the proper viscosity for the selected method of application. Most paints are supplied in the proper viscosity for brush or roller application. For conventional spraying a somewhat lower viscosity is desired, and some thinning usually is required.

Thinners are added to paints before spraying to accommodate for solvent loss during travel time between the gun and surface. Evaporation time can be controlled only to a limited extent by varying the spraying distance, yet some operators have tried to increase the evaporation time of slow solvents by holding the gun as far as 20 in. or more from the surface—with consequent poor spray results.

Within each class of solvent—aliphatic and aromatic hydrocarbons, ketones, alcohols, esters, etc.—a selection of solvents is available having a wide range of evaporation rates. The drying time of paints is adjusted properly by selecting faster or slower evaporating solvents for thinning, as required. The evaporation rate for a number of solvents is presented in Table 5-2, all determined under the same empirical set of conditions.

In thinning any paint for spraying, the solvent first must be selected from the proper class of thinner for the paint being applied. The selection is then made based on evaporation rates which will result in a full wet coat being applied, which flows out properly without runs when the gun is held at a normal distance from the work—say 8 to 10 in.

The greater the volatile content of the paint, the lower will be the

TABLE 5-2 Relative Solvent Evaporation Rates
Values determined under the same controlled conditions (Based on an evaporation rate of carbon tetrachloride of 1 min)

Solvent	Evap. time, min	Solvent	Evap. time, min
Aliphatic hydrocarbons		Aromatic hydrocarbons	
Hexane	0.7	Toluene	2.7
Heptane	1.6	Xylene	10.8
Kerosine	325	Aromatic high-flash C.T. naphtha	30
Mineral spirits	35–37	Heavy aromatic naphtha	580
Naphtha, VM and P	4.0–4.5		
Ketones		Ethers	
Acetone	0.8	Ethylene glycol ethyl ether	29
Cyclohexanone	25	Ethylene glycol methyl ether	12
Isophorone	200	Isopropyl ether	0.7
Methyl ethyl ketone	1.3–1.4	Ethylene glycol butyl ether	58
Methyl isobutyl ketone	3.6–3.7	Diethylene glycol ethyl ether	600+
Alcohols		Miscellaneous	
n-Butyl alcohol	13	Butyl acetate	5.8
Ethyl alcohol (95%)	3.4	Ethylene glycol ethyl ether acetate	29
Isopropyl alcohol (91%)	3.6	Ethylene dichloride	1.3
Methyl alcohol	1.6	Trichloroethylene	1.3
Diacetone alcohol	29	Turpentine	14
		Nitropropane	5.3

Note: Refer to Table 3-4, page 70, for trade names of the glycol ether solvents.

SOURCE: The Solvents and Chemicals Companies: Central Solvents and Chemicals Co., Chicago, Ill.

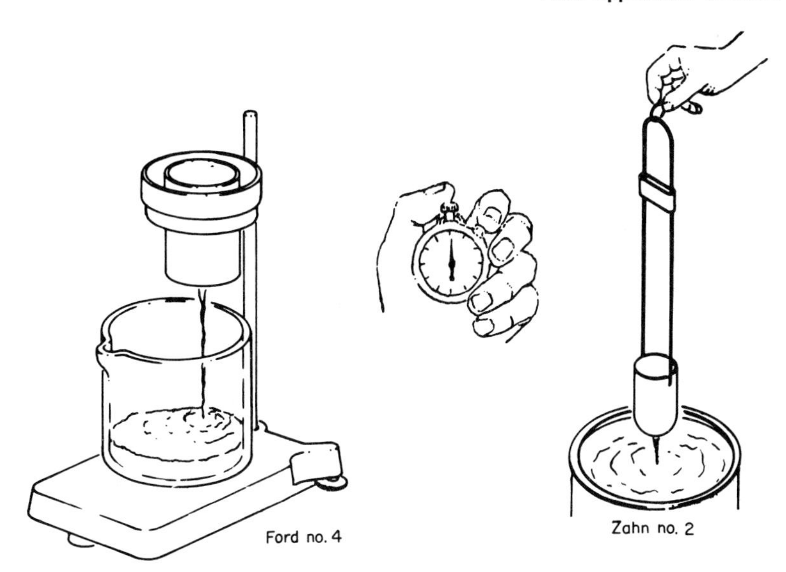

FIG. 5-1 Illustrations of cup viscosity determinations.

build per coat. If the total dry film thickness of a coating system is below the specified minimum, the coating will exhibit a shorter life than expected. Therefore, thinners always should be used sparingly.

Viscosity considerations: The viscosity of a paint is a very important factor in selecting the proper equipment for spray application. Viscosity is a measure of a liquid's resistance to flow and may be determined by various methods. Table 5–3 presents the relationship between viscosities as determined by the more common methods.

For painters, the most commonly employed method for determining viscosity depends upon timing the flow of a measured amount of paint through a known orifice. The usual containers used for this purpose are the No. 2 Zahn cup and the No. 4 Ford cup, and the flow time is measured in seconds by a stopwatch.

The two main factors which affect paint viscosity are the temperature and the amount of solvent contained in the paint. Since temperature has an appreciable effect, the viscosity is determined and reported at a known temperature, for example: 25 sec No. 2 Zahn at 77°F.

The effect of temperature on viscosity of a representative paint fluid is presented in Fig. 5–2, where the viscosity is measured by one of the cup methods.

Temperature variations during a normal working day may cause wide fluctuations in viscosity. Paint which has been stored in a cold place should be allowed to warm up before use. When warming paint in a hot water bath or electrically, the lid should be loosened and left ajar. Pref-

TABLE 5-3 Viscosity Conversion Chart

(For fluids at 77°F without unusual thixotropic properties)

	Very thin (Solvents, water)		Thin (Lacquers, sealers)			Medium (Primers, varnishes)					Heavy (Oil paints, enamels, latexes)							Very heavy (Filled coatings, mastics)								
Centipoise	10	20	30	40	50	60	70	80	90	100	120	150	170	200	250	300	350	400	450	500	1,000	2,500	5,000	7,500	10,000	15,000
Brookfield CPS	10	20	30	40	50	60	70	80	90	100	120	150	170	250	250	300	350	400	450	500	1,000	2,500	5,000	7,500	10,000	15,000
Fisher No. 1 S × D	20	30	39	50																						
Fisher No. 2 S × D	...	15	18	21	24	29	33	39	44	50	62															
Ford No. 3 S × D	...	12	19	25	29	33	36	41	45	50	58	70														
Ford No. 4 S × D	5	10	14	18	22	25	28	31	32	34	41	47	52	58	67	74										
Gardner-Holdt Bubble Units S × D	A-4	A-3	A-1	A	...	B	...	C	...	D	E	F	G	H	J	L	N	P	Q	S	W	Z-1	Z-3	Z-4	Z-5	Z-6
Krebs Unit S × D	...	...	...	...	30	33	35	37	38	40	43	47	49	52	57	60	62	64	66	68	85	114	140			
Parlin No. 7 S × D	27	32	43	50	57	64																				
Parlin No. 10 S × D	11	13	15	16	17	18	20	22	23	25	30	35	40	45												
Parlin No. 15 S × D	...	...	...	...	...	...	...	...	...	...	...	...	...	10	...	15	...	20	...	25	47	135	232	348	465	697
Parlin No. 20 S × D	...	...	...	...	...	...	...	...	...	...	...	...	...	...	...	...	...	...	...	8	17	55	83	125	167	250
Parlin No. 30 S × D	...	...	...	...	...	...	...	...	...	...	...	...	...	...	...	...	...	...	...	...	...	...	19	29	38	58
Saybolt (Universal) (SSU) S × D	60	100	160	210	260	320	370	430	480	530	580	740	845	1,000	1,240	1,330	1,475	1,950	2,215	2,480	4,600	11,600	23,500	35,000	46,500	69,500
Stormer (150 Gr.) S	...	10	12	15	16	18	20	22	25	27	32	38	44	49	...	...	...	...	...	114	223	450	1,090	1,635	2,180	
Zahn No. 1 S × D	30	37	44	52	60	68																				
Zahn No. 2 S × D	16	18	20	22	24	27	30	34	37	41	49	62	70	82												
Zahn No. 3 S × D	...	...	...	...	...	...	...	...	10	12	14	17	19	23	29	34	40	46	51	57						
Zahn No. 4 S × D	...	...	...	...	...	...	...	...	...	10	11	13	15	17	21	24	27	30	34	37						
Zahn No. 5 S × D	...	...	...	...	...	...	...	...	...	...	...	...	...	10	13	15	18	20	22	25	49					
Sears Craftsman Cup S × D	...	...	...	19	20	21	23	24	26	27	31	36	39	44												

Note:

CPS = Centipoise

S = Viscosity readings in seconds

D = Specific gravity

S × D = Time in seconds × specific gravity of fluid

SOURCE: Binks Manufacturing Co., Chicago, Ill.

erably the temperature of the paint should be at least as high as that of the surface to which it is applied. However, the paint should not be heated over 90°F for conventional application methods—spraying, brushing, or roller coating. Temperatures employed in hot spraying are discussed in a later section of this chapter.

General considerations in applying paint: Good painting practice requires planning. For major jobs a time schedule should be prepared which will reflect the progress of surface preparation and painting. Proper allowances should be made for such items as erection and moving of scaffolding, drying times, expected weather conditions, temperature of air and surface to be painted, and sunshine and shade during the working hours. The pot life of the mixed material of catalyzed coatings is a very important factor and must be considered in the work schedule. Allowance also should be made for other construction and erection activities going on at the same time which may cause damage or contamination of the painted surfaces.

The drying of intermediate coats should be checked by a competent person, because drying varies with temperature, humidity, and ventilation conditions. The drying times presented for selected paints in the "Specifications" supplement to this manual will serve as a guide, but they cannot be expected to apply fully under all conditions.

Before painting, the contacting surfaces of joining parts, and areas which are inaccessible after assembly, should be sealed with a mastic material. This is particularly important for contacting surfaces of dis-

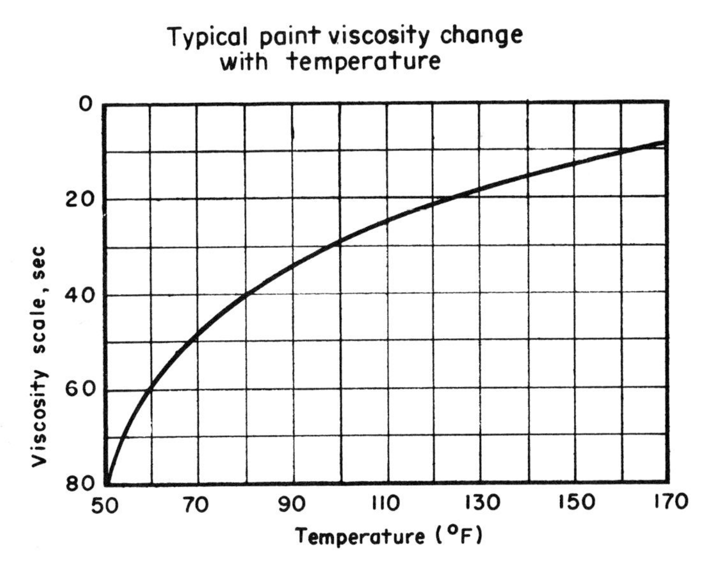

FIG. 5-2 Typical viscosity versus temperature curve.

similar materials, such as different metals, metals joining nonmetallic surfaces, wood in contact with masonry, etc.

Most paints will draw back from sharp edges, leaving a very thin film on the edge where corrosion will start first. To compensate for this, an additional strip coat of primer is recommended to be applied to all edges. This may be done either before or after applying the first prime coat to all surfaces and edges.

In general, paints should be applied only to surfaces that are thoroughly dry and under such conditions of humidity and temperature as will promote evaporation rather than condensation. This means that the temperature of the surface should be above the dew point of the air. A practical test is to wipe a damp cloth on the metal surface to be painted. If the moist streak dries within a few minutes, the drying conditions should be satisfactory for paint application.

Strong hot sunlight falling on surfaces being painted or the application of paint to hot surfaces may cause the paint to "boil," resulting in porosity, cratering, or blistering. Oil binders may be partly decomposed by strong sunlight during application, forming organic acids which promote corrosion of metals. The paint will not dry properly and may remain tacky for a long time when this occurs.

Wind or strong drafts will cause application difficulties because of too rapid evaporation of solvents. The painter will have too short a lap time when brushing. When spraying, the amount of overspray and the danger of dusting are increased. Also, dust or sand will be deposited on the freshly painted surface, spoiling the appearance of glossy finishes and forming points where corrosion may start.

Skilled workmen should be employed in painting. Inexperience produces unsatisfactory results and leads to unreasonable wastage of material and effort. Therefore it pays to train painters who are lacking in skill in order to acquaint them with the most efficient working methods and with modern paint materials and equipment. Good labor relations are important to the quality as well as to the volume of work produced.

When painting jobs of major importance are put out to contract, the crew should be known to be well skilled or should be given an intensive short training course. This may prove more economical and even more effective than full-time supervision.

The most conspicuous "trademark" of good workmanship is a neat appearance of the work: absence of ridges, sags, runs, drops, or laps; no unnecessary brush marks; and a general cleanliness of the working area. Orderliness on the job, personal cleanliness, and care of tools and materials are essential to healthy and safe working conditions.

The most important criteria for skillful painting, however, are not so conspicuous: conscientious surface preparation, thorough mixing of the paint, limited use of thinner, uniform film thickness, conservation of paint materials, proper drying and protection of paint coats, systematic and expeditious work habits, and a well-organized supply of materials, tools, and equipment at the job site.

Spray Application of Paint

Introduction: All spray systems are related in that the paint fluid is first "atomized," or broken up into small droplets before it is applied to the surface to be coated. The energy to accomplish this atomization may be provided by any of three sources, leading to the recognition of three basic methods of paint spraying:

1. Atomization by intersecting jets of compressed air: "conventional" or "air atomization" spray systems.
2. Atomization through the sudden release of high pressure as the fluid is ejected through a small orifice: "airless" or "hydraulic" spray systems;
3. Atomization through the application of high voltage static electricity as the paint flows off a sharp edge or point, or after it has been preatomized by either of the above two methods: "electrostatic" spray systems.

Each of these methods of spray application of paint exhibits certain advantages, disadvantages, and limitations either in use or in results. The proper choice of spray method will follow from an understanding of these characteristics, as presented below.

Air Spray. With air spraying the compressed air must supply the energy required to atomize the paint. It takes about a barrel of air to atomize a pint of paint. Air atomization produces the highest quality surface finish because the finest degree of atomization is obtained. This is the most versatile method of spraying: spray rates range from small touch-up painting to full production work, and paint viscosities can be handled from heavy semisolids to the thinnest of lacquers.

Air spray produces the greatest amount of offspray or paint fog, however. The air used to atomize the paint bounces off the surface, carrying a considerable portion of the paint particles with it. Dry particles of paint from this fog will fall back onto the surface beyond the spray pattern, resulting in a dusty or sandy appearance. Also, the fog may drift away for some distance to coat another surface such as an automobile, with unfortunate results.

Airless Spray. Airless spray painting is most suitable for large-volume applications of maintenance paints and protective coatings, where the fineness of finish is not of paramount importance. Airless spray permits the application of heavy film thicknesses with the minimum number of strokes, or the application of coatings to the greatest area in the least amount of time.

With no atomizing air to rebound, paint fog is greatly reduced over air spraying. Furthermore, the spray pattern usually is more confined than with air atomization, so less overspray occurs—"overspray" being the paint which falls outside of the area being painted.

Airless spray does not produce the fine degree of atomization required in fine finish work. However, modern airless spray equipment will successfully apply decorative and maintenance paints to walls and structural surfaces with fully satisfactory results.

Electrostatic Spray. Applying paint by means of electrically charged particles is accomplished with a considerable reduction in paint loss from fog and overspray. This is due to the strong electrostatic attraction which draws the particles to the surface being painted.

Hand-held electrostatic guns are available whereby the electrostatic charge is applied to paint particles already atomized by air spray or airless methods, or where the electrostatic charge is applied to the paint as it flows off the sharp edge of a rotating concave disk or bell. The air and airless electrostatic guns are more versatile in use, since they also may be employed as conventional air spray or airless guns when so desired.

These electrostatic spray methods exhibit generally the same advantages and disadvantages of conventional spray application, but with the one important added advantage of electrostatic attraction of the paint particle. This attraction is so strong that paint particles which otherwise would miss the article entirely are drawn back to coat the side or back surface—an effect termed "wraparound."

Electrostatic spraying thus finds its greatest application to painting intricately shaped articles or those consisting of combinations of tubes, rods, and grills, where considerable paint loss due to overspray otherwise would occur.

Heating the paint before application results in certain advantages common to all three basic spray methods discussed above. By reducing the paint viscosity through heat instead of by thinning, thicker films may be applied per coat without porosity from solvent evaporation; dry time between coats is shortened; solvent loss is greatly reduced; and rebound is less with air spray because lower air pressures are required for atomization. Hot spray painting is discussed in greater detail in the sections covering air spray and airless spray methods.

Conventional air spray application: Approximately 70 percent of all paint spraying is by air atomization. This undoubtedly is due to the broad versatility of this method, which permits the application of the thinnest stains to the heaviest-bodied coatings over a wide range of application rates.

A broad selection of spray guns has been designed to meet all types of application requirements, ranging from the conventional application of paints, coatings, mastics, and filled materials, to special effects from mother of pearl finishes, flocks, glass beading, etc. Some 1,500 combinations of spray nozzles are available which were developed to meet these many requirements.

The application of conventional maintenance coatings and decorative finishes by air spraying involves relatively few choices in equipment, however. These are concerned with choice of paint-feed method (siphon or pressure feed, by pressure tanks or pumps), and choice of atomization (by external or internal mix nozzles). These choices will be discussed in greater detail in the following section.

Air Spray Gun and Nozzles. The basic functions of an air-atomizing spray gun are to use compressed air to break up the paint into small droplets, and to give these small droplets direction. The spray gun itself provides two convenient valves to start and stop the flow of compressed air and fluid. The mixing of air and paint takes place outside the spray gun between the "horns" of the air nozzle in "external mix atomization." The mixing of air and paint takes place inside the air nozzle of the spray gun when an "internal mix nozzle" is employed.

Paint can be brought to the spray gun by creating a vacuum through use of a "siphon" type of external mix nozzle. This vacuum permits atmospheric pressure to force the paint through a tube leading from an open-top container to the nozzle of the spray gun. This method of atomization and paint delivery is called "siphon spraying." When heavier fluids or higher production rates are required, the paint is forced to the nozzle under positive pressure. This method is called "pressure-feed" spraying.

Air Nozzles. There are two basic types of air nozzles of interest to the spray painter: external mix nozzles and internal mix nozzles, illustrated in Fig. 5-4.

The most common type of air nozzle used today is the external mix nozzle, due to its ability to atomize the widest range of materials and to produce the finest degree of atomization for the highest quality finish. With external mix nozzles, the paint and atomizing air do not come into contact with one another until they have left their respective nozzles. Atomization occurs external to the nozzle—see Fig. 5-5.

There are two types of external mix nozzles: siphon-feed and pres-

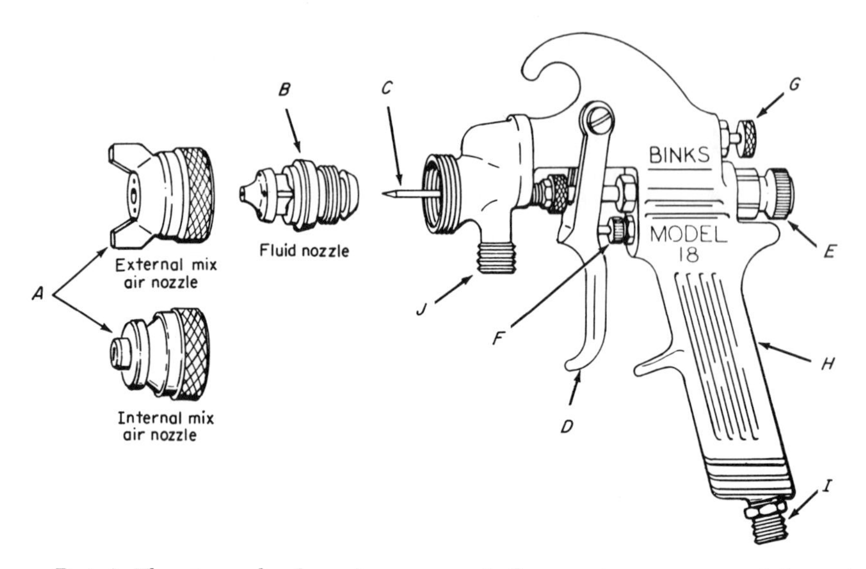

Part *A*. The air nozzle of an air spray gun is the most important part of the entire spray gun. The air nozzle directs the air jets which atomize the fluid and give the particles direction and velocity.

Part *B*. The fluid nozzle is the second most important part of a spray gun, since this part provides a control over paint flow rates.

Part *C*. Needle assembly (in conjunction with the fluid nozzle) acts as a stop-start valve for the paint flow through the spray gun; and in siphon-feed systems only, regulates the flow of paint.

Part *D*. Trigger operates and controls the air and fluid valves.

Part *E*. Fluid control assembly provides mechanical pressure on the needle valve so that it will close when the trigger is released.

Part *F*. The air valve controls the airflow rate to the air nozzle.

Part *G*. Side port control or fan control regulates the spray-pattern width by controlling the air supply to the "horns" on the external mix air nozzle.

Part *H*. The spray gun body handle is designed to hold conveniently all of these parts and to give the operator a balanced and comfortable handle for spraying.

Part *I*. Air inlet, to provide a connecting point for the air hose, normally ¼ in. NPS(male).

Part *J*. Fluid inlet, to provide a connecting point for a siphon cup or the fluid hose, normally ⅜ in. NPS(male).

FIG. 5-3 The principal components of an air spray gun.

sure-feed nozzles. All siphon-feed nozzles are external mix nozzles. With siphon-feed nozzles the stream of compressed air creates a vacuum in the paint container, allowing atmospheric pressure to force the paint from the attached container to the spray head of the gun. This gun is usually limited to one-qt containers or smaller, and is easily identified by the fluid nozzle protruding slightly beyond the air nozzle face. Siphon-feed guns are used in limited spraying such as touch-up work, or where there are many color changes and small amounts of material are involved.

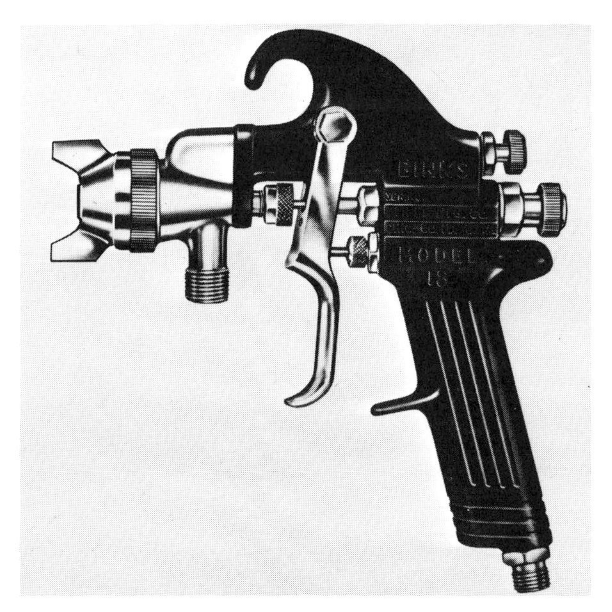

FIG. 5-3 (Continued)

Most production-type paint spray guns are equipped with pressure-feed external mix air nozzles. Nozzles of this type are not designed to create a vacuum. The fluid tip usually is flush with the air nozzle face. Paint is forced to the gun by positive pressure from a pressure cup, tank, or pump. When large amounts of paint of the same color are being used, when the material is too heavy to be siphoned from a cup or container by suction, or when fast application is required, a pressure feed gun is used.

With external mix nozzles the air jets atomize the paint stream, and shape and direct the flow of paint particles into the desired spray pattern. The spray pattern size is determined by the design of the air nozzle, the method of feeding the spray gun, and the cohesive nature

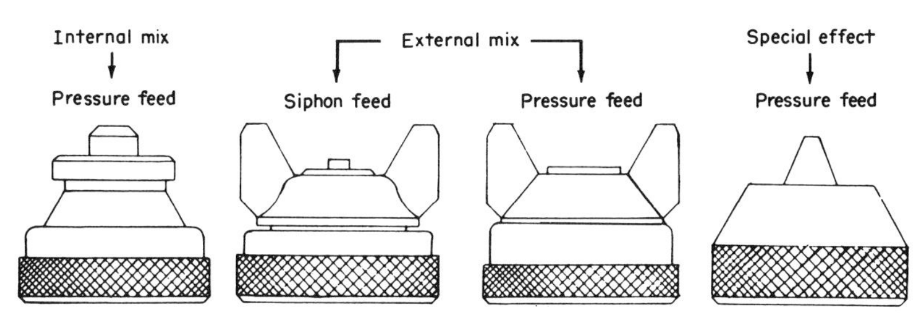

FIG. 5-4 The basic types of air spray nozzles.

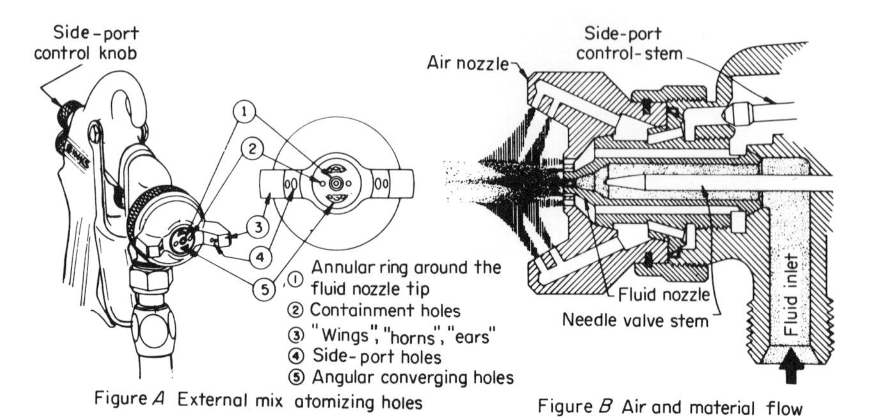

FIG. 5-5 Typical external mix nozzle construction.

of the paint being sprayed. The spray pattern width and elongation can be adjusted by controlling the air to the side port jet. Turn the control knob until the desired pattern is achieved. For maximum efficiency of atomization the fluid volume should be controlled by adjusting the fluid pressure or by changing the size of the fluid nozzle orifice, and not by attempting to control the flow by the fluid needle valve.

The internal mix nozzle is often employed in high production applications of maintenance paints and protective coatings where fine finish is not required. However, this type of nozzle is not recommended to be used with highly abrasive or coarsely pigmented coatings because of nozzle tip wear, and also is not recommended with catalytically cured coatings or certain fast-dry paints because of the tendency of these materials to plug the exit orifice.

In an internal mix air nozzle, air and fluid are mixed in a cavity inside the air nozzle before being released to the atmosphere. These air nozzles have either a round hole or a slot from which the material sprays. When the atomized paint leaves the air nozzle, it will expand and conform to the shape of the air nozzle opening.

The internal mix air nozzle can be used only with pressure-feed systems. Lower air and fluid pressures are required than with external mix nozzles, and these pressures must be regulated so as to be equal at the nozzle. Because of the low air pressure used, this type of nozzle does not give as fine a degree of atomization as does an external mix nozzle. Furthermore, less air volume is required than with external mix nozzles. This, coupled with the lower air pressures, means that considerably less overspray and rebound will occur.

Disadvantages of the internal mix air nozzle are few but important:

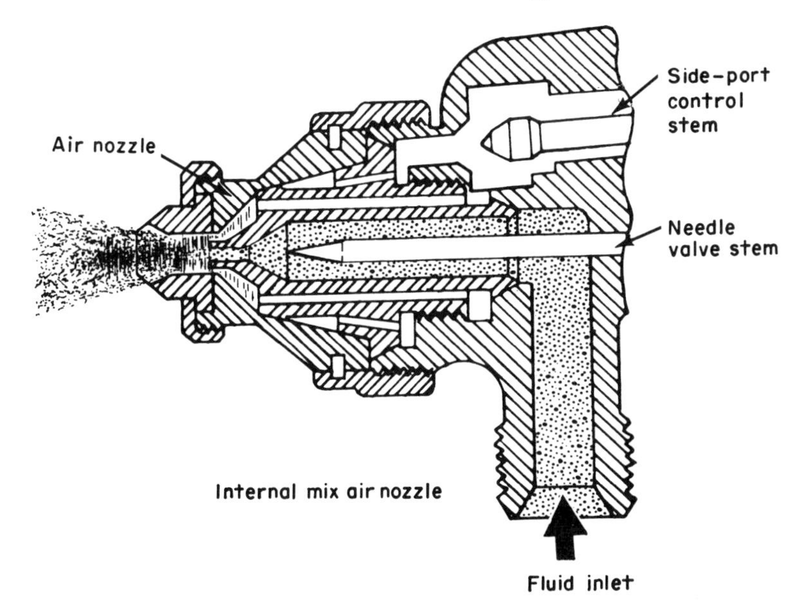

FIG. 5-6 Typical internal mix nozzle construction.

the spray pattern size and shape cannot be controlled except to a limited degree by varying air and fluid pressures; a relatively coarse atomization is produced, not suitable for fine finishes; nozzles are subject to considerable wear (although replacement is easy and low in cost); and certain types of paints, such as fast driers, will clog the exit slot or hole.

A summary of the relative merits and limitations of external and internal mix nozzles is presented below:

External mix air nozzle	*Internal mix air nozzle*
Advantages	*Limitations*
Fine atomization	Coarse atomization
Ready control of spray pattern	Spray pattern size fixed
No wear on air nozzle	Nozzle tip subject to wear
Siphon and pressure feed types	Pressure feed only
Limitations	*Advantages*
Larger air rates required	Lower air rates needed
High air pressures required	Lower air pressures used
Maximum overspray and rebound	Minimum overspray and rebound
Limited film build per pass	Maximum film build per pass
Limited size spray pattern	Maximum size spray pattern

Fluid Nozzles. The functions of the fluid nozzle are to meter the paint fluid, and to direct the paint stream into the air jets. The fluid nozzle also forms a seat for the fluid needle, to shut off fluid flow.

Factors affecting selection of fluid nozzles are viscosity, pigments and fillers (coarse, fibrous materials require large nozzle sizes), and fluid flow rates required for proper atomization. Fluid nozzles are constructed of hardened steel, nitralloy for abrasive use, stainless steel to prevent corrosion, and with carbide inserts for extreme abrasion.

The velocity of paint flow through the nozzle must be considered in selecting the proper nozzle. If the fluid nozzle orifice is too small and high pressure must be used to obtain the desired volume of flow, then inefficiency results. This is because the fluid stream emerges at too great a velocity for the air jets to properly atomize the fluid. Paint pressure should not exceed 18 psig at the gun inlet.

The following table presents the usual size range of fluid nozzles to be considered when spraying various fluids. Equipment manufacturers should be consulted for more detailed information.

Orifice range, in.	Fluidity	Examples
0.022–0.040	Very thin	Solvents, water
0.040–0.052	Thin	Lacquers, sealers
0.052–0.070	Medium	Primers, varnishes
0.086–0.110	Heavy	Oil paints, latexes
0.125–0.500	Very heavy	Filled coatings, mastics

The wide range of paint flow rates obtainable through proper selection of nozzle size, without exceeding the maximum recommended 18-psig pressure, is illustrated for a thin viscosity paint (20 sec, No. 4 Ford cup) in Table 5-4:

TABLE 5-4 Air Spray Fluid Nozzle: Flow Rate vs. Fluid Pressure

Fluid nozzle orifice size, in.	Delivery, oz/min		
	8 psig	12 psig	18 psig
0.040	12.5	15.0	19.0
0.046	17.0	21.0	27.0
0.052	20.5	25.5	33.0
0.059	24.0	30.0	40.0
0.070	28.0	36.0	48.0

It is recommended to select a nozzle size which is sufficiently large to pass enough material for the rate of spraying desired, while using the lowest practical fluid pressure. It should be remembered that the important consideration is to have a low velocity stream of paint emerging from the fluid nozzle.

For maximum efficiency of atomization, the fluid volume should be controlled by adjusting the fluid pressure or by changing the size of the fluid nozzle orifice, rather than by attempting to control the flow by the fluid needle valve.

The determination of the volume of fluid delivered in ounces per minute is simple. Merely free-flow the paint into a graduated beaker for a measured period of time, and convert this volume and time into ounces per minute.

Paint-feed Equipment. Paint may be pumped to the gun or fed to the nozzle by siphon or under positive pressure using paint containers ranging in size from a few ounces to 60 gal or larger. The production sprayer employs pumps or pressure tanks (5- and 10-gal sizes are common) for area painting, and pressure or siphon cups for touch-up work.

Pressure tanks conserve air supply since the compressed air only applies pressure to the tank, while paint pumps require compressed air for operation. Pumps are more convient and portable in use, especially when changing from one paint system to another. A representative paint pump for air spraying would have a 2:1 fluid/air pressure ratio, with a pumping capacity of 2 to 3 gpm and a free air demand of 4 to 5 cfm. Other low pressure pumps are available with pumping ratios up to 10:1 and pumping capacities up to 10 gpm or more. The larger pumps require as much as 30 cfm free air (pressured to 75 to 100 psig). Typical low-pressure paint pumps are presented in Fig. 5-7.

The air supply must have sufficient capacity to meet the requirements of both the gun air nozzle and the pump, when one is used. Air nozzle requirements depend upon the capacity of the nozzle and its type—whether external or internal mix. Nozzle air requirements are indicated in the following table for a few representative production-type nozzles.

A review of Table 5-5 will show that for the generally applied coatings, the air compressor must have a capacity of around 15 cfm (free air) for each gun employed. An additional 15 to 25 cfm air is required if a paint pump is used, plus 5 cfm for each air-supplied mask. As an estimate of compressor capacity, 6 to 7 cfm free air will be compressed to 50 psig per electric horsepower. Refer to Table 4-3 on page 100 for compressor capacities at 100-psig discharge pressures.

An air regulator must be employed to control the air pressure, and an extractor is required to filter out moisture, oil, and other impurities from the air. The regulator and extractor often are combined in one unit. The extractor should be placed at least 25 ft from the compressor to allow the compressed air to cool before it reaches the oil and water

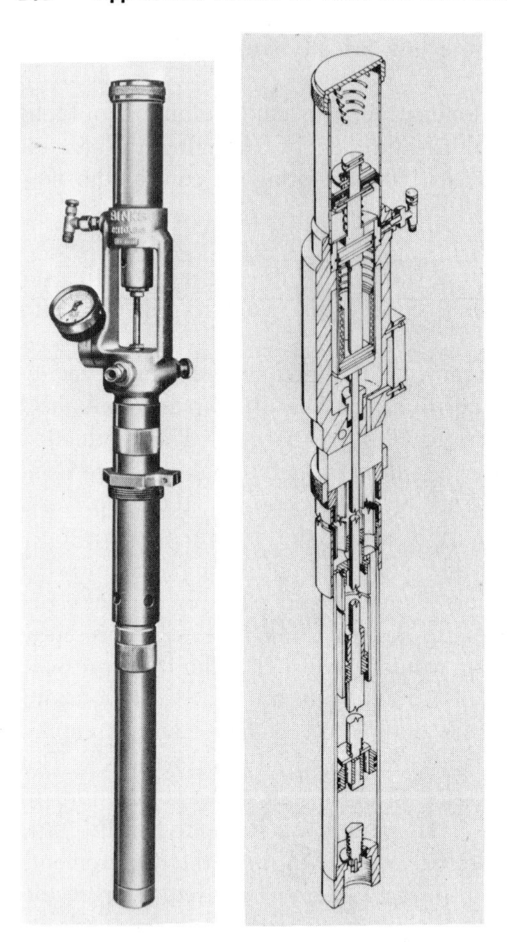

FIG. 5-7 A low-pressure paint pump. (*Binks Manufacturing Co., Chicago, Illinois.*)

TABLE 5-5 Representative Nozzle Air Requirements

Fluidity of material applied	Free air demand, cfm, at indicated gun air pressures				
	Internal mix		External mix		
	30 psig	50 psig	30 psig	50 psig	70 psig
Very thin: solvents, stains, wash primers	2.2–3.1	3.0–5.2	4.5–5.1	7.5–8.7	10.0–12.2
Thin: lacquers, sealers, lubricants	3.1–3.9	5.2–5.5	9.0–10.0	14.2–15.0	19.0–20.0
Medium: primers, varnishes, lacquers	3.1–3.9	5.2–5.5	9.0–10.1	14.3–15.1	20.0–20.1
Heavy: oil paints, enamels, latexes	4.6–8.6	6.8–13.2	9.0–10.0	14.3–15.3	20.0–20.8
Very heavy: mastics, filled coatings	8.6–13.1	13.2–20.3	10.0–10.6	15.3–16.5	20.8–22.3

extractor. The air line should be sloped back towards the compressor, so that condensed moisture will flow back into the air receiver where it can be removed through the drain cock.

The air hose from the air regulator to the gun should not be less than 5⁄16-in. ID if no more than 50 ft long. For greater lengths a larger size hose should be used. The air pressure should be set to allow for the pressure drop in the hose, and is adjusted while the trigger is pulled. The air hose pressure drops presented in Table 5-6 may be used in estimating this loss in pressure.

Fluid hose must be selected to withstand the solvent action of the paint. Air hose also can be used as the fluid hose for water-base paints. Most solvent-base paints and protective coatings can be handled by the ordinary "solvent resistant" fluid hose which has a Thiokol-blend inner tube. Lacquers, particularly if heated, require the use of nylon tubing or a special hose with a polyvinyl alcohol resin inner tube.

The fluid hose must be sized for the viscosity of the material sprayed, so that the pressure loss in the hose is acceptably low. For normal paints, 3⁄8-in. ID hose is used in lengths up to 50 ft and 1⁄2-in. hose for longer lengths. Heavy-bodied coatings require the use of 1⁄2-in. and 3⁄4-in. hoses.

TABLE 5-6 Air Pressure Drop in Air Hose
(At airflow rates required by an air nozzle having an air capacity of 12 cfm (free air) at 50 psig)

Hose size	Input pressure, psig	Pressure drop through hose, psig		
		10-ft length	25-ft length	50-ft length
1⁄4 in.	40	8.0	12.8	24.0
	50	10.0	16.0	28.0
	60	12.5	19.0	31.0
	70	14.5	22.5	34.0
	80	16.5	25.5	37.0
	90	18.8	29.0	39.5
5⁄16 in.	40	2.8	4.0	8.5
	50	3.5	5.0	10.0
	60	4.5	6.0	11.5
	70	5.3	7.3	13.0
	80	6.3	8.8	14.5
	90	7.5	10.5	16.0
3⁄8 in.	40	1.1	1.6	3.3
	50	1.4	2.1	4.1
	60	1.8	2.6	4.9
	70	2.3	3.3	5.9
	80	2.9	4.1	6.8
	90	3.7	5.1	7.8

SOURCE: Binks Manufacturing Co., Chicago, Ill.

Representative equipment systems are presented below. The system recommended will depend upon the amount of paint to be sprayed, its viscosity, and the air supply available.

Air Spraying of Heated Materials. Hot spraying was developed primarily in order to reduce the amount of solvents required for thinning a paint to application consistency. In hot spraying, the viscosity is lowered instead by warming the paint. Heating to around 120°F markedly reduces the viscosity, but above 120 to 130°F the effect of temperature increase is less—see Fig. 5-2 on page 129.

As the heated paint leaves the gun nozzle, it cools rapidly to ambient temperature because of the cooling effect of the expanding atomizing

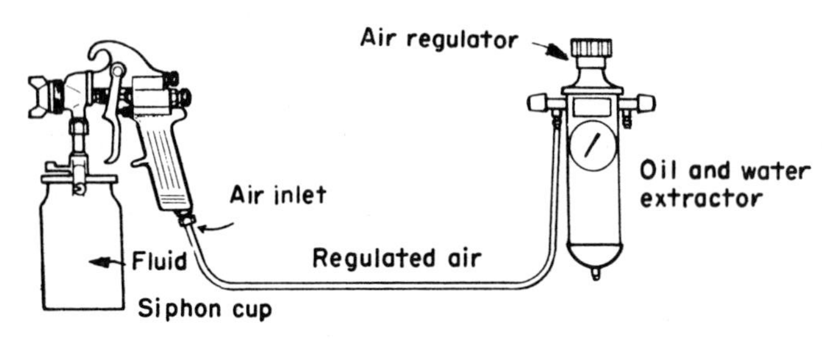

SIPHON CUP HOOKUP
This method is suitable for limited spraying and touch-up work. Atomizing air is regulated at extractor. Paint flow and atomization are adjusted by fluid control screw on gun, and by regulation of atomizing air pressure.

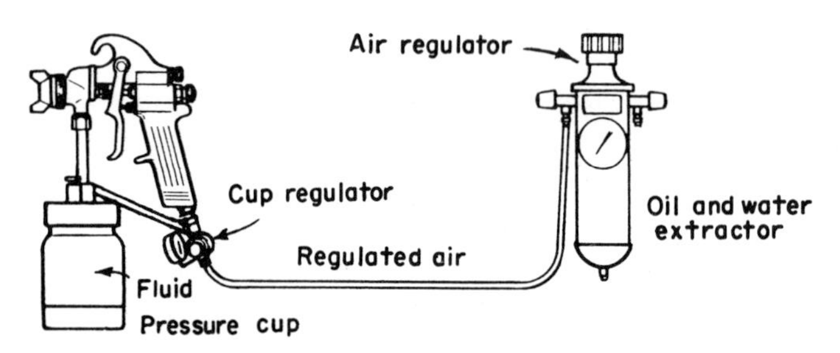

PRESSURE CUP HOOKUP
This method is ideal for fine finishing with limited paint spraying. Atomizing air is regulated at extractor, and fluid pressure by the cup regulator. Atomizing air passes through the cup regulator.
Pressure cups also are available without a cup regulator, in which case fluid pressure equals atomizing air pressure. For heavy fluids and internal mix nozzle spraying, fluid flow rate is adjusted by the fluid control knob on the gun.

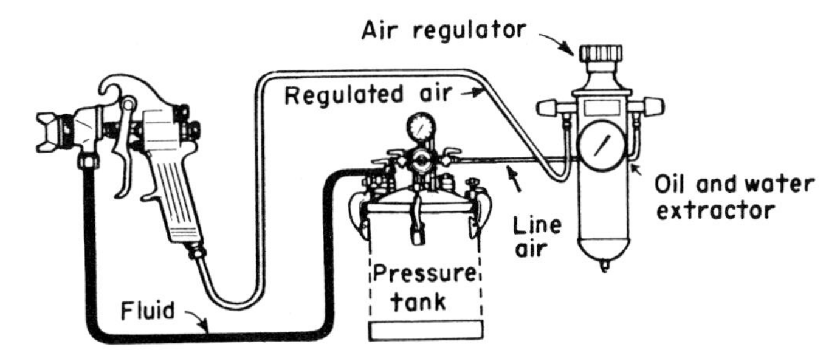

PRESSURE TANK HOOKUP
This system is ideal for medium-production spraying, especially when compressed air capacity is limited. Atomizing air is regulated at extractor, and fluid pressure at the tank regulator. Pressure tanks also are available with both tank and atomizing air regulator located on the tank.

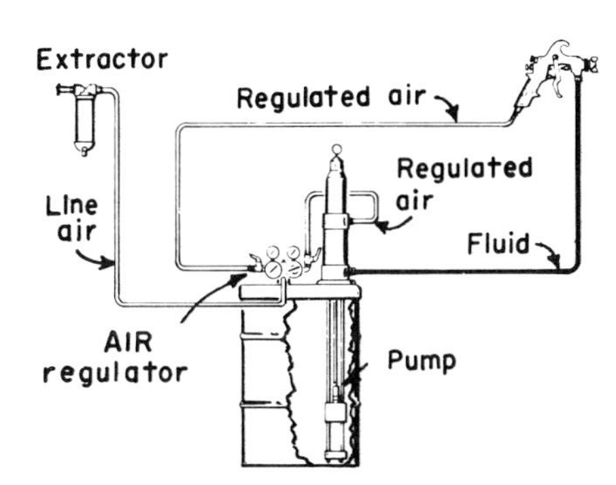

PAINT PUMP HOOKUP
This system is recommended for full production-rate spraying. Pumps are available for mounting on 5- to 55-gal containers. Atomizing air and pump-operating air pressures are controlled at the dual regulator usually located on the container cover. Wall-mounted pumps which "siphon" paint from the container through hose also are available.

air. Paint at 120°F cools rapidly to 77°F when 3 in., to 70°F when 6 in., and to 68°F when 9 in. from the nozzle. The paint is therefore considerably more viscous when it meets the surface than when it was atomized at 120°F.

The advantages in hot spraying when applied to either air or airless spray methods are important. These are:

Costs are reduced because of less solvent loss

Heavier films may be applied per pass and per coat without pinholing due to solvent evaporation

Faster dry times between coats permit quicker recoating

Viscosity of the material is controlled through temperature control, and remains constant throughout the day or season

Better flow-out on the surface (with some types of coatings)

With air spray, less fog and rebound occur because lower air pressures are needed for atomization

With airless spray, lower fluid pressures are required for atomization.

The few limitations in painting by hot spraying are equally important:

Not all paints or coatings will withstand heating—for instance rapidly setting catalytically cured coatings

Additional equipment is required: a paint heater, and a second fluid hose to the gun in recirculating systems

Temperatures must be watched and kept below the boiling range of the solvents.

Two types of hot spray systems are in common use: the circulating system, where the gun draws off paint as needed from a constantly circulating supply of heated paint; and a dead-end system, where a single fluid hose to the gun carries heated paint. A special gun with a circulating head is required with circulating systems.

Several types of low pressure paint heaters are available which will heat paint to 170°F, employing electrically heated coils or a hot water heat exchanger. The dead-end hot spray system makes use of conventional equipment but with a heater added in the paint feed line to the gun.

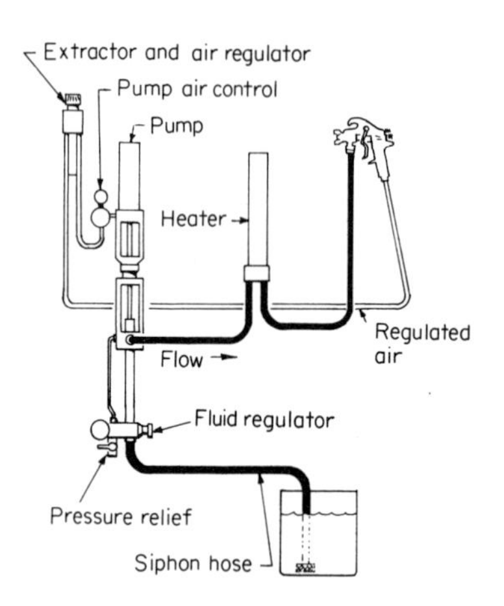

Heated siphon feed dead-end system

DEAD-END HOT AIR SPRAY SYSTEM

This system is suitable when spraying is sufficiently continuous to maintain hot paint in the hose. Equipment is simple: only requires a heater to be added to otherwise conventional equipment.

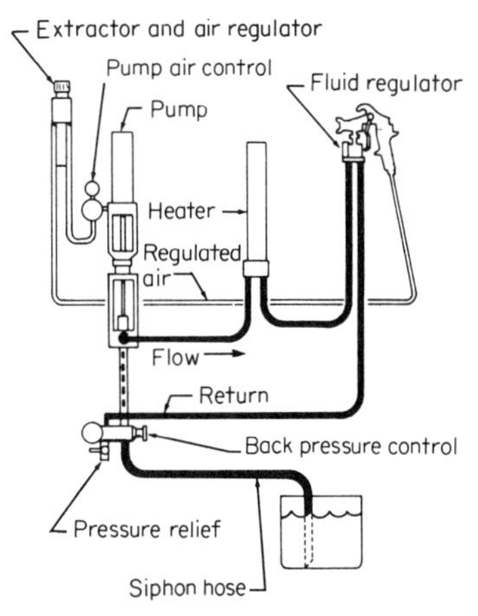

CIRCULATING HOT AIR SPRAY SYSTEM
This system is suitable for intermittent spraying conditions and for in-plant painting.
A gun with a circulating head is required, connected to paint feed and return hoses.

Care and Operation of Equipment. The majority of troubles experienced with spray painting are directly traceable to neglect or improper cleaning of the equipment after use. The spray gun is a precision tool built to fine tolerances. It must be kept lubricated and clean:

1. Do not immerse the entire gun in solvent. This allows sludge and dirt to collect in the air passages, which later will be blown on the surface being sprayed. Also, solvents remove lubricants from the gun and dry out packings.

2. Lubricate the spray gun with a drop of light machine oil each day. Lubrication points are shown in Fig. 5-8.

Note: Never use oils or lubricants containing silicones.

3. Wash off the gun body with a rag dipped in solvent.

Lubrication Points in an Air Spray Gun:
1. Trigger pivot
2. Fluid packing nut
3. Air valve packing
4. Fluid needle spring
5. Side port and control knob

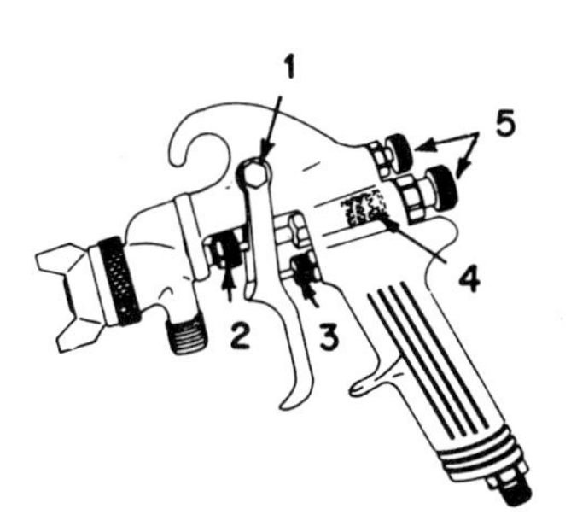

FIG. 5-8 Lubrication points in an air spray gun.

To clean a spray gun equipped with a siphon cup, thinner or suitable solvent should be siphoned through the spray gun by inserting the cup tube into an open-top container of that solvent. Move trigger constantly to thoroughly flush the passageway and clean the tip of the needle.

To clean a spray gun used with a pressure tank, open the vent on the tank and loosen the air nozzle. Hold a piece of cloth wadded in the hand over the air nozzle and pull the trigger. The air will back up through the fluid nozzle and force the fluid in the hose back into the tank (sometimes referred to as "blowing back"). After emptying the tank, put in enough clean thinner or solvent to wash the interior of the hose and spray gun thoroughly, and spray this through the spray gun until it runs clean. *Caution:* never remove the tank lid before releasing the air pressure first.

When pumps are used, drain paint from the hose and gun back into the paint container by disconnecting the fluid hose from the pump discharge connection. "Blowing back" will help drain paint from the hose. Connect the pump suction to a supply of solvent and circulate solvent through pump, hose, and gun until clean.

To clean the gun exterior surfaces, immerse only the front end or spray head portion of the spray gun until solvent just covers the fluid inlet connection. Use a hair bristle brush to remove accumulated paint. If the air nozzle holes are clogged, remove the air nozzle and wash parts in clean solvent. A pipe cleaner, match, or broom bristle will help to clean the holes. *Never* use a hard or sharp instrument or piece of wire for this purpose.

When an air spray gun becomes dirty or out of adjustment, the consequence will be leaking air or paint, or improper operation which results in faulty spray patterns. A review of the more frequent spray gun operating problems and their probable causes is presented below:

1. Fluid leaking from the needle packing nut is caused by a loose packing nut or dried-out packing. Tighten the nut until leaking stops, or replace packing if required
2. Air leaking from the front of the gun may be due to:
   a. Foreign matter on the air valve stem or seat
   b. Broken air valve spring
   c. Packing nut too tight
   d. Bent air valve stem
   e. Air valve assembly loose, or gasket leaking
   f. The gun is a bleeder-type spray gun, where air leakage is normal
3. Fluid leaking from the front of the gun may be due to:
   a. Worn or damaged needle or fluid nozzle
   b. Dirt in fluid needle seat
   c. Fluid packing nut too tight
   d. Broken fluid control spring
   e. Wrong size needle

4. A jerky or fluttering pattern is caused by:
   a. Air leaking into the fluid line or passageway
   b. Lack of paint
   c. Loose or cracked fluid siphon tube
   d. Loose fluid nozzle
   e. Loose fluid packing nut or worn packing

Before spraying is started, the fluid pressure first is adjusted with the atomizing air turned off, by triggering the gun and observing the solid stream of paint flowing from the fluid nozzle. With the fluid control knob on the gun turned fully open and the gun held in a horizontal position, the pressure is adjusted so that the paint stream begins to bend downward at a distance of about 3 ft from the nozzle. Remember, for paints of normal viscosity, this pressure is not to exceed 18 psig at the gun inlet fitting. If higher pressures are required a larger size fluid orifice should be used. Always use the fluid nozzle which, with the lowest possible fluid pressure, still provides the paint delivery desired.

The atomizing air pressure is now adjusted to the lowest pressure required for proper atomization. To start, 25-psig air pressure can be tried and adjusted upwards in 10-lb test increments until proper atomization is obtained. (With thin lacquers, stains, etc., air pressure should start at 10 psig.) The fan-width control knob is adjusted to give the desired spray pattern shape after proper pressures have been determined.

Faulty air spray patterns reviewed below are readily recognized and corrected:

Cause

Dried material in side port (A) restricts passage of air through port on this side. Full pressure of air from clean side port pushes fan pattern in direction of clogged side.

Correction

Dissolve material in side port with thinner. Do not probe in any air nozzle openings with metal probes or wires.

Cause

Dried material around the outside of the fluid nozzle tip at position (B) restricts the passage of atomizing air over a portion of the center ring opening of the air nozzle. This faulty pattern can also be caused by a loose air nozzle, or a bent fluid nozzle or needle tip.

Correction

If dried material is causing the trouble, remove air nozzle and wipe off fluid tip, using rag wet with thinner. Tighten air nozzle. Replace fluid nozzle or needle if bent.

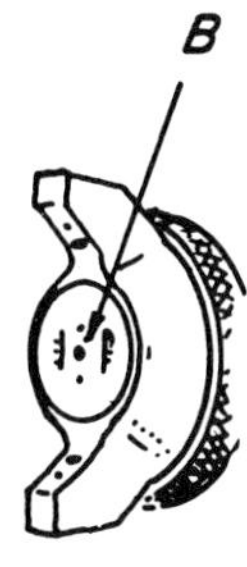

Cause

A split spray pattern (heavy on each end of a fan pattern and weak in the middle) is usually caused by (1) atomizing air pressure too high, (2) attempting to get too wide a spray with thin paint, (3) not enough paint available—fluid pressure too low.

Correction

(1) Reduce air pressure. (2) Open fluid control (D) to full position by turning to left. At the same time turn spray width adjustment (C) to right. This reduces width of spray but will correct split spray pattern. (3) Increase fluid pressure.

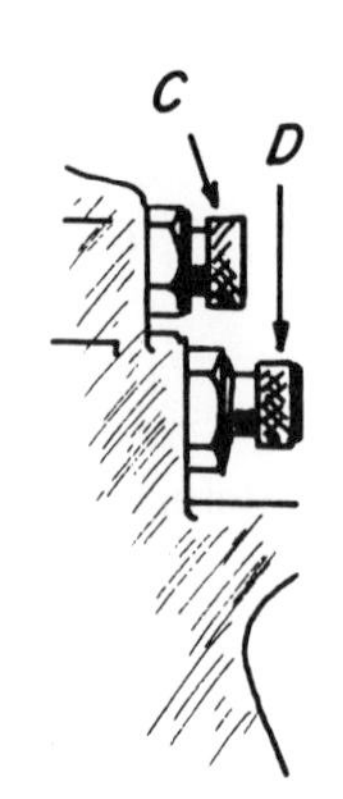

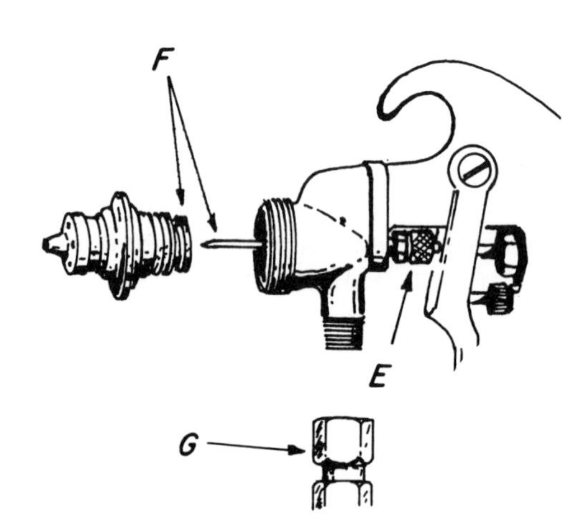

Cause

Air entering the fluid supply causing spitting: (1) Dried packing or missing packing around the fluid needle valve, which permits air to get into fluid passageway. (2) Dirt between the fluid nozzle seat and body, or a loosely installed fluid nozzle. (3) A loose or defective swivel nut on the siphon cup or hose fluid.

Correction

(1) Make certain that all fittings and connections are tight. (2) Back up knurled nut (E), place two drops of machine oil on packing, replace nut, and finger tighten. In aggravated cases, replace packing. (3) Remove fluid nozzle (F), clean back of nozzle and nozzle seat in gun body using rag wet with thinner, replace nozzle and draw up tightly against body, using the proper tool. (4) Tighten or replace swivel nut (G).

Cause

A fan spray pattern that is heavy in the middle or a pattern that has a coarse "salt-and-pepper" effect indicates that the atomizing air pressure is not sufficiently high, or that there is too much paint being fed to the gun.

Correction

Increase pressure from air supply. Reduce the fluid pressure or install a smaller fluid nozzle.

An acceptable paint spray job can be obtained even when following poor application practices. However, the paint work will be appreciably more costly than if proper spraying techniques had been followed. Since gun handling techniques apply equally well to both air and airless spraying methods, they are presented in some detail following the discussion on airless spraying—see page 157. However, the fundamentals of good air spraying are repeated in outline form below:

1. Use the lowest possible air and fluid pressure when operating an air spray gun. Select a fluid nozzle so that the fluid pressure is under 18 psig at the gun.
2. Operate the gun with the fluid control adjustment wide open.
3. Adjust the fan width to be suitable for the job.
4. Spray from the proper distance (from 6 to 10 in.), and keep this distance constant.
5. Hold the spray gun perpendicular to the work throughout the spray stroke.
6. Move the spray gun parallel to the work surface.
7. Move the spray gun at a speed which assures a full wet coat on the work surface; always move the gun at this same speed.
8. Feather the ends of spray strokes by "triggering"; always trigger while the gun is in motion.
9. Lap strokes at least 50 percent for finish work; keep the laps uniform in width.
10. Feather the spray stroke overlaps in the middle of a panel by triggering and using a short arcing motion.
11. Follow the recommended pattern of strokes for the job.
12. Cover all hard-to-reach areas; band edges and corners.
13. Keep clothing and equipment from touching wet painted surfaces.
14. Work as a team when possible.

Airless spray application: Airless or hydraulic spraying is a method of paint application that does not use compressed air to atomize the paint or coating material. Hydraulic pressure alone is used to atomize the fluid by discharging it at high pressure (400 to 4,500 psig) through a small orifice in the spray nozzle. As the fluid is suddenly released at these pressures it breaks up into small droplets resulting in a fine atomized spray. The fluid is discharged at such a high velocity that after atomization, sufficient momentum remains to carry the minute particles to the surface being painted.

The fluid pressure required for proper atomizing depends primarily upon the viscosity of the material being applied, and also to some extent upon its cohesive nature. Very thin fluids such as stains and lacquers are successfully atomized at pressures around 400 to 800 psig. The usual paint and protective coating requires 1,500 to 2,000 psig; some heavy paints such as the water-base PVA paints need 2,500 psig; while atomizing pressures for many mastics go above 3,000 psig and even at these pressures fine atomization is not achieved.

Airless spraying is a high-production method for applying decorative maintenance paints and protective coatings. Coatings can be applied to surfaces as fast as the painter can move the gun. The degree of atomization with airless spray is not sufficiently fine, however, for fine finish work, and the spray pattern edges are not as finely feathered as with air atomization. Both the fineness of atomization and spray pattern edge feathering are improved by use of a preorifice which can be inserted in the gun between the valve and nozzle—see page 152.

Airless spray painting is cleaner and faster than conventional spraying methods. Airless enables the user to handle many painting jobs that are impractical by conventional spraying methods.

A primary advantage of the airless spray method is that it greatly reduces fog (or rebound), often to less than half of that experienced with a conventional air-atomized system. This makes the use of high production spray equipment possible in places and in applications where paint otherwise would be more slowly applied by brush or roller coater.

Another significant advantage is the ability to apply heavy coating thicknesses when desired. Heavy bodied paints with the high viscosity of 50 sec No. 4 Zahn cup, for example, can be applied to metal surfaces producing a film thickness of 3 mils (dry) in a single coat.

When high production rates are required, a continuous spray from ½ to 1½ gal per min can be applied. Moreover, many coatings can be sprayed in their as-received unthinned state by properly adjusting the spraying pressure.

Airless Spray Gun and Nozzles. The equipment employed in airless spraying is much less complex than that required for air spraying. Airless spray equipment essentially consists of a high pressure pump with proper pressure and volume capacities, a source of compressed air to operate the pump, high pressure fluid hose, and the airless gun.

The airless spray gun, itself, is far simpler than an air spray gun. In essence, the gun is merely a convenient means of holding and operating the fluid nozzle and valve.

The airless nozzle determines both the volume which can be sprayed and the spray pattern width. In selecting the nozzle it is necessary first to consider the fluid viscosity and application rates desired, in order to determine the proper nozzle orifice size. And for each orifice size a selection of nozzles is available providing fan widths from round to wide angle.

The quantity of fluid sprayed is determined by the size of the orifice; the film thickness applied is determined both by the orifice size and the fan angle. Two nozzles having the same orifice size but different spray angles will deposit the same amount of paint, but over a different area. Note that orifices are not circular but are elliptical in shape; the diameters referred to are equivalent to a circular diameter having the same flow capacity. A good rule is to determine the largest fan angle and the smallest orifice that is practical for the specific fluid and application method.

Since the spraying characteristics—volume and spray pattern—are fixed with each fluid orifice, a selection of nozzles is required to meet a variety of spraying conditions. Although over 90 types of airless spray nozzles are available, for most applications only two nozzles are usually needed: a nozzle with a large orifice (0.018 to 0.021 in.) and a wide

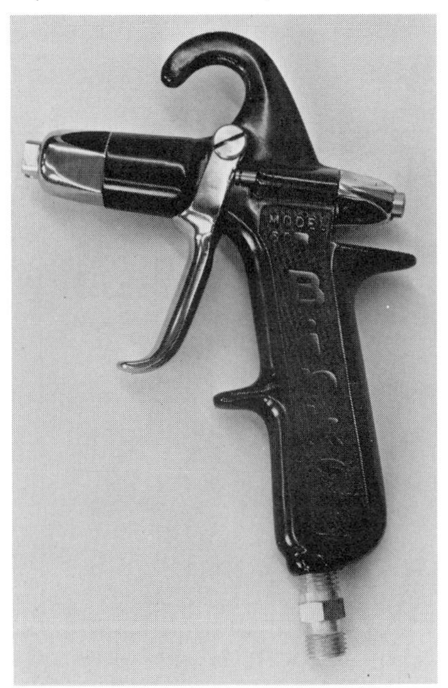

FIG. 5-9 A typical airless spray gun.

angle (60 to 80°), and one with a small orifice (0.015 in.) and a narrow spray angle (30°). For application of the usual house paint and maintenance coating, the most commonly employed nozzle has an 0.018-in. orifice and a 60° fan.

The following table may serve as a guide in choosing airless spray nozzles:

Orifice range, in.	Fluidity	Examples
0.007–0.010	Very thin	Solvents, water
0.011–0.013	Thin	Lacquers, sealers
0.013–0.018	Medium	Primers, varnishes
0.015–0.031	Heavy	Oil paints, latexes
0.026–0.072	Very heavy	Filled coatings, mastics

A chart presenting a few representative airless nozzle flow rates for various fluids at different pressures is given in Table 5-7.

Some paints such as the latex paints are difficult to atomize hydraulically. With these, finer atomization is produced by inserting a secondary orifice between the valve and nozzle. This "preorifice" serves to greatly increase turbulence in the fluid stream, which aids in the fluid breakdown into particles as it emerges from the nozzle orifice. The

TABLE 5-7 Airless Nozzle Flow Chart
(Delivery, oz/min)

	Orifice size, in.	500 psig	1,000 psig	1,500 psig	2,000 psig	2,500 psig
Water........	0.007	4	5	6	6.7	7
	0.009	4.5	5.7	6.8	8.4	10
	0.011	6.5	8.5	12	14	15
Lacquer.......	0.013	12	15	19	22	26
	0.015	13	19	24	27	32
	0.016	14	22	29	34	39
Latex paint....	0.018	....	19	26	35	44
	0.021	....	24	32	46	56
Plastisol......	0.026	....	21	34	51	65
	0.031	....	26	48	65	85
	0.036	....	32	68	95	126
Coal-tar epoxy	0.043	....	....	61	105	143
	0.072	....	....	112	151	190

Note: There are variables in equipment and fluids which cause this chart to be approximate; however, it is sufficiently accurate for proper equipment selection.
SOURCE: Binks Manufacturing Co., Chicago, Ill.

preorifice size should be selected to be close to the nozzle size, but never smaller.

Airless Pump and Hose. Two factors must be considered in selecting an airless pump: the fluid pressure required, and the volume to be pumped at these pressures. Although the usual airless pump is a high pressure air-driven pump, the electrically operated airless pump is becoming increasingly popular.

Most electric airless pumps now available have sufficient capacity to meet the spray requirements for full production rate applications of paint at pressures up to 3,000 psig. The typical pump is driven by an explosion-proof 115 volt electric motor having an electric current demand sufficiently low to be handled by an average house circuit.

The principal advantages of electric airless spray pumps over conventional air-driven pumps are those of convenience:

No need for a source of compressed air—an air compressor and air hose are no longer necessary.

Maximum portability and convenience—compact, light in weight, maneuverable, powered at any electrical outlet.

Very quiet in operation—no compressor or pump air exhaust noise.

These conveniences result in considerable savings in time and labor

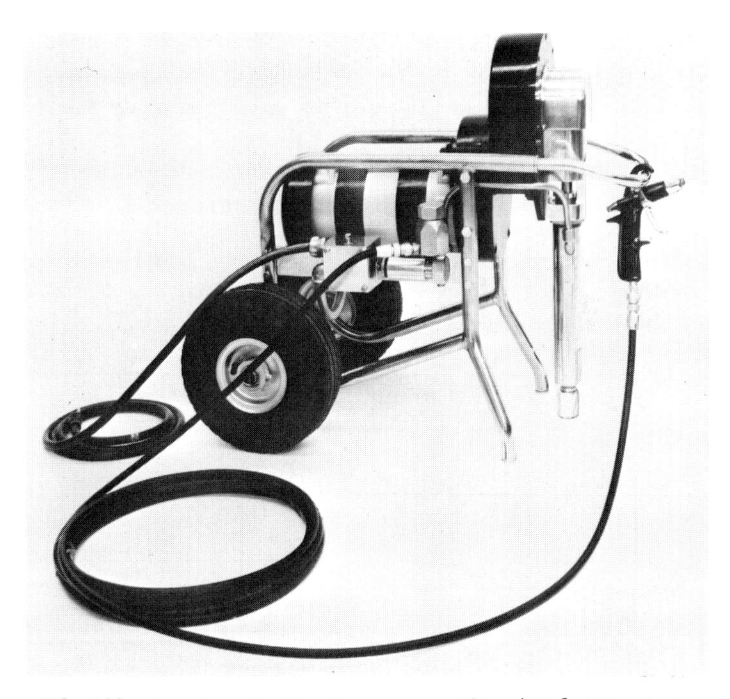

FIG. 5-10 A modern electric airless spray outfit. (*Binks Manufacturing Co., Chicago, Illinois.*)

to set up the equipment and start paint application. Furthermore, the quiet operation of an electric airless pump permits the economical airless spray application of paint in locations where otherwise paint must be applied by brush or roller—for example, in occupied buildings, schools, hospitals, and other institutions.

The air-driven airless spray pump is similar to the pump used for air spraying as discussed on page 139, except for use of a larger air motor in relation to the fluid section. The fluid pressures required for airless atomization are varied by controlling the air pressure to the pump. The maximum fluid pressure obtainable will depend both upon the size of the air motor, and the air pressure used to operate the pump.

The pump pressure rating is presented as a ratio—for instance, 25:1—which indicates the approximate ratio of air piston area to effective fluid piston area. Consequently, a 25:1 pump driven by 100 psig air would be expected to provide approximately 2,500 psig fluid pressure. However, under operating conditions valve and seal efficiency, friction, etc. reduce this output pressure by 20 to 40 percent, and this must be taken into account in selecting an airless pump.

The pumping capacity of a reciprocating pump depends upon the fluid cylinder size and pump stroke rate. The pump capacity presented in manufacturers' literature thus is meaningful only if the corresponding stroke rate also is given. The maximum recommended stroke rate for most pumps is 120 strokes per min, or 60 cycles per min. The air demand to operate the pump is, of course, directly related to the stroke rate.

A typical pump for the airless spray application of maintenance paints is rated as a 30:1 pump with a 1¼-gpm capacity. The actual relation-

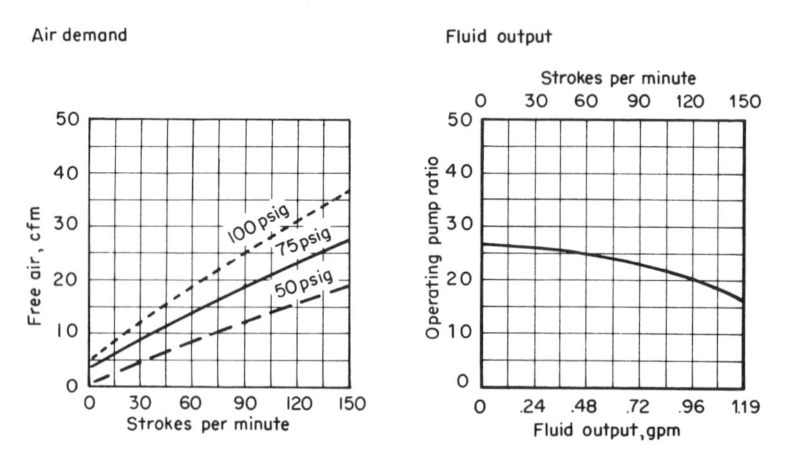

FIG. 5-11 Pump performance curves for a typical airless spray pump.

TABLE 5-8 Pump Air Demand Related to Output
(Pump free air demand, cfm)

Pumped volume, oz/min	Paint discharge pressure, psig				
	1,000	1,500	2,000	2,500	3,000
5	0.7	0.8	1.0	1.3	
10	1.1	1.3	2.2	2.7	3.5
15	2.0	2.5	3.0	4.0	6.0
25	3.5	4.0	5.0	6.0	10.0
50	6	8	10	13	19
100	12	15	20	25	34
150	17	23	30	37	
200	22	27	37	45	
300	35	40	55		
400	50				

SOURCE: Binks Manufacturing Co., Chicago, Ill.

ship between the pump ratio, stroke rate, pumped volume, and air demand for this pump is presented on two charts in Fig. 5-11.

A wide range of pumps for airless spray are available to meet all operating needs, having pump ratios ranging up to 60:1 or 70:1 and output volumes up to 4 or 5 gpm. Since one gun with an 0.018-in. orifice has a normal paint handling capacity of around ¼ gpm, several guns can be operated from one pump—which represents a savings in equipment costs.

The air required for pump operation may be considerable, and must be considered in selecting a pump for airless spray painting. The approximate pump free air demand, as related to fluid output volume and pressure, is presented in Table 5-8. This table is based upon the test results of a number of types of airless pumps under actual operating conditions.

The most commonly used airless fluid hose is a high pressure nylon hose which contains a static grounding wire to ground the static electricity generated at the nozzle. This hose is sufficiently solvent resistant for all types of maintenance paints and protective coatings commonly applied. In the 3⁄16- and ¼-in. sizes usually employed, this hose exhibits a burst strength of 8,000 to 10,000 psig.

Airless Spraying of Heated Materials. Airless spraying of heated paints provides much the same benefits that are obtained with hot air spraying, which are discussed on page 142. However, a particularly important additional benefit derived from heating paint materials for airless spraying is the reduction of viscosity and paint cohesive forces.

This permits the use of considerably lower fluid pressures for atomization with a finer degree of atomization than when spraying unheated paints.

Equipment systems for hot airless spraying are similar to those described for air spraying of heated materials with the exception that high pressure airless equipment is employed.

Operation of Airless Spray Equipment. Airless spray equipment, because of the high pressures involved, must be maintained and operated with special attention to certain important details. These special precautions are reviewed below:

1. Always be sure that all fluid connections are tight before starting the pump, and that the gun is effectively grounded through connections and hose.
2. Make certain that the high pressure fluid hose has not been damaged by kinking, bending, or abrasion.
3. Never use standard pressure fittings but only high pressure fittings when modifying the paint distribution system.
4. Never carelessly handle the spray gun, and never direct its spray toward any part of the body—especially when the nozzle has been removed. The high pressure fluid jet from a gun with the nozzle removed can cause serious injury.
5. Never attempt to change the nozzle without first either disengaging the trigger or relieving the fluid pressure.
6. Never attempt to disconnect the hose or gun without first relieving the fluid pressure.

The principal operating problem associated with airless spraying is that of nozzle plugging. Since nozzle openings are minute, paint feed to the gun must first be passed through filters with openings not much larger than the nozzle orifice. Since filters usually are located at the pump discharge, deposits in the hose still can cause plugging.

To save the time required to remove and clean out a plugged nozzle, a nozzle holder is available as a gun attachment by which the nozzle position can be reversed. By rotating the nozzle 180° it can be back-pressured and usually blown clean.

To reduce the frequency of nozzle plugging, it is essential that all equipment be thoroughly cleaned by solvent flushing after use. This applies particularly to the fluid hose; dried paint flaking off the inner surface of used paint hose is the most common cause of plugging.

A review of the more common airless gun operating problems and their causes is presented below:

1. Gun spits—the flow of atomized paint from the nozzle is not continuous. This may be due to:
   a. Air entering the fluid system because of loose connections in the pump suction line

b. Dirt in the gun or in the feed line to the gun
c. Nozzle screen dirty
d. Valve closing mechanism out of adjustment
e. Valve seat chipped or cracked
f. Valve stem packing too tight, or dry and sticking

2. Gun will not shut off. This may be caused by:
   a. Dirt in the gun holding the valve open
   b. Valve parts worn, or the valve seat chipped or cracked
   c. Valve closing mechanism badly out of adjustment
   d. Valve stem packing much too tight, or dry
3. Gun does not spray any fluid. This may be due to:
   a. No paint being fed to the gun because of plugging in the pump suction, or loss of air pressure to the pump
   b. Paint filters plugged solid with deposits
   c. Broken valve mechanism so that the trigger cannot open valve

Faulty airless spray patterns are as readily recognizable as with air spray application, but they are due to different causes. Some of the more common faulty patterns and their probable causes and correction are reviewed on page 158.

Spray gun handling techniques: The proper techniques for handling and operating a spray gun in order to obtain a smoothly applied film of paint of desired thickness are essentially the same whether the atomization is by air spray or airless methods.

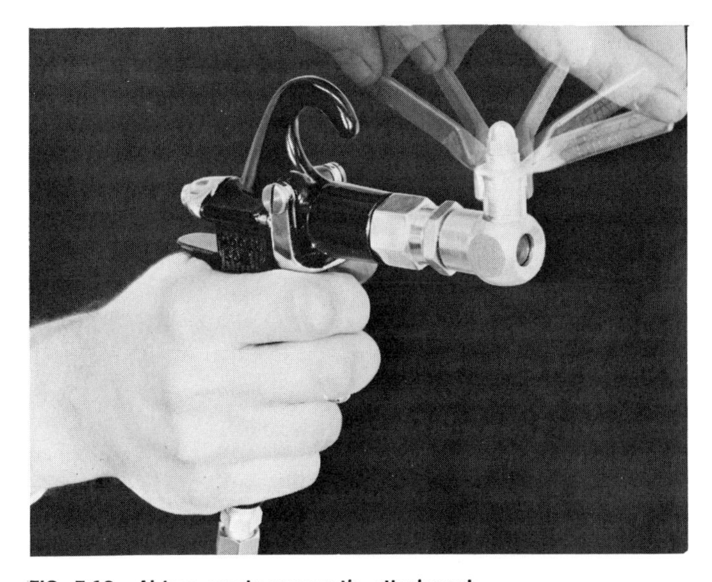

FIG. 5-12 Airless nozzle reverse-tip attachment.

Problem	Causes	Correction
PATTERN HAS TAILS	(1) Fluid not atomizing properly due to too low a pressure, too low a velocity through the orifice, or insufficient volume of paint. (2) Fluid too viscous or cohesive for satisfactory atomization.	(1) Increase fluid pressure. (2) Change to smaller nozzle orifice size. (3) Reduce fluid viscosity. (4) Clean gun and filter(s). (5) Reduce number of guns using pump. (6) Install properly matched preorifice spray insert.
PATTERN HEAVY IN CENTER	(1) Worn nozzle. (2) Fluid will not spray by airless method.	(1) Corrections same as above. (2) Change to air atomizing spray system.
DISTORTED PATTERN	(1) Plugged or worn nozzle.	(1) Clean or replace nozzle.
PATTERN EXPANDING AND CONTRACTING (SURGE)	(1) Pulsating fluid delivery. (2) Insufficient air supply to pump. (3) Leak in suction tube or line. (4) Pump capacity too low. (5) Material too viscous.	(1) Change to a smaller nozzle orifice size. (2) Install pulsation chamber in system, or drain existing one. (3) Reduce number of guns fed by the pump. (4) Increase air supply to air motor. (5) Remove restrictions in system. Clean or remove screens or filters; use larger hose or pump if necessary. (6) Inspect siphon tube and hose assembly for leak. (7) Reduce fluid viscosity.
ROUND PATTERN	(1) Worn nozzle. (2) Fluid too heavy for nozzle size. (3) Fluid will not spray by airless method.	(1) Replace worn nozzle, or change to a larger size. (2) Increase fluid pressure. (3) Thin material. (4) Install preorifice spray insert. (5) Change to air-atomizing spray system.

Good spray painting results are accomplished through adherence to the following proper procedures and motions:

1. Hold the gun perpendicular to the surface and always at the same distance.
2. Always move the gun across the surface at the same steady rate.
3. Always lap the same distance when making successive strokes.
4. Trigger each stroke in the same manner.

These essential techniques will be presented in greater detail in the following discussion. Although an air spray gun is used in the illustration, the discussion applies equally well to airless application techniques.

The spray gun stroke is made by moving the gun parallel to the work and at a right angle to the surface. The distance from the gun to work should be from 6 to 10 in. for the air spray of most paints (and up to 14 in. for acrylics). Airless spray requires a greater distance: 10 to 15 in.

The closer the gun is held to the work, the more paint is deposited on the surface and the faster the gun must be moved to prevent sags. Holding the gun too far from the work causes dry spray and excessive spray dust, particularly with air spray application.

The speed of the stroke is an important factor not only in the conservation of material but also in conservation of operator energy. The proper speed is one that will allow a full wet coat to be applied with each stroke. This wet coat should be of a thickness just under that thickness at which a sag will occur. The thickness at which materials sag is not always the same, as this is a function of many factors such as temperature, the type of solvents being used, and the viscosity of the paint.

It is important to hold the gun at all times perpendicular to the surface being sprayed. The obvious faults with arcing the stroke are shown in Fig. 5-13. "Heeling" and "toeing" the gun, which is tipping the gun forward and backward during the stroke, results in coating streaks of uneven thickness and also causes excessive spray fog and dusting.

Proper lapping—the distance between successive spray strokes—is essential in producing a uniform film thickness. If the spray pattern is 12 in. wide and the gun is moved downward 6 in. before making the next spray stroke, the paint is being applied with 50 percent lapping—see Fig. 5-14.

No greater lapping should be used than is absolutely necessary to give the finish uniformity desired. For fine finish work, 50 to 70 percent lapping may be necessary; however, a 25 percent lap (recoating 3 in. in a 12-in. pattern) is sufficient for most heavily applied maintenance coatings.

Work with straight uniform strokes spraying alternately from left to right and right to left, holding stroke speed, distance, and lapping as uniform as possible.

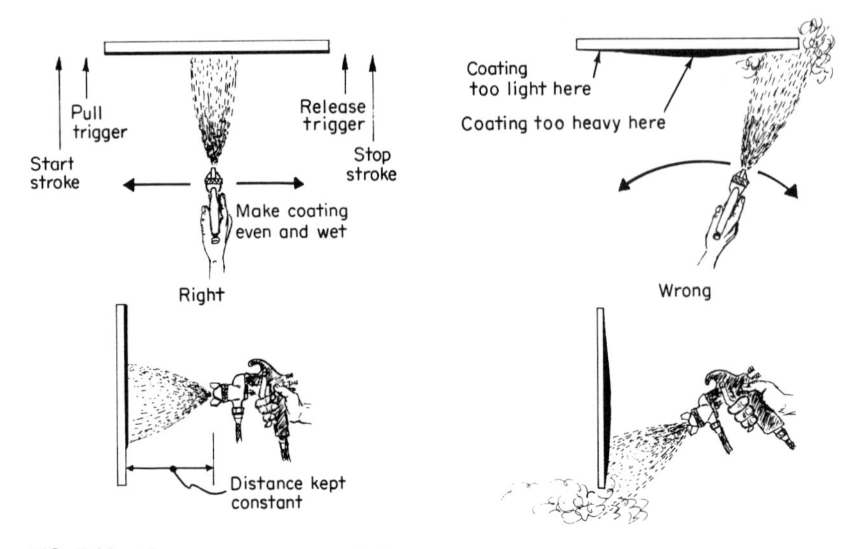

FIG. 5-13 The proper spray gun stroke.

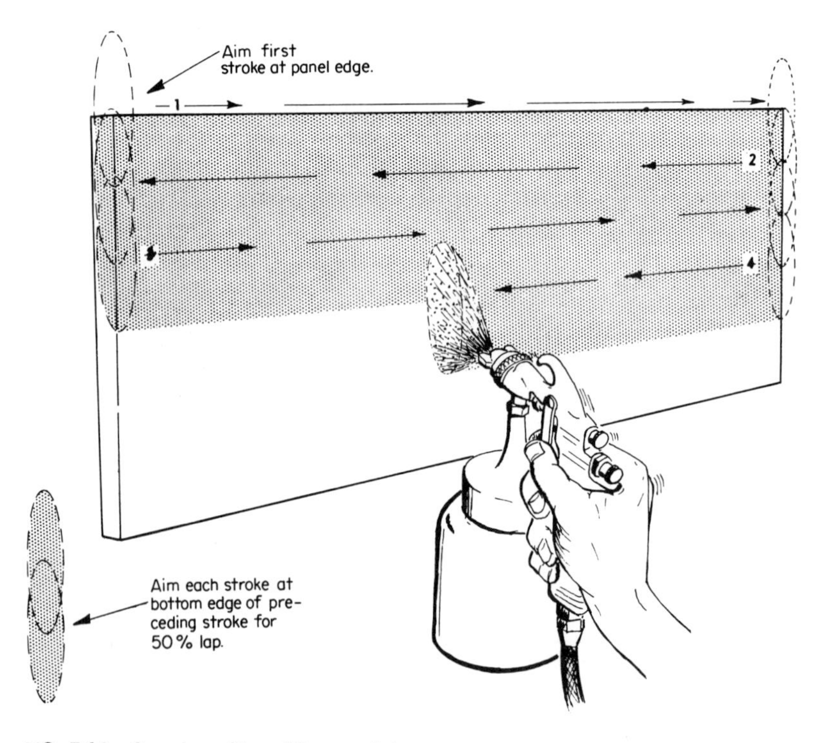

FIG. 5-14 Spraying with a 50 percent lap.

The trigger controls the action of the gun; proper triggering is the heart of correct spraying technique. Triggering serves two important purposes: the needle valve actuation cleans the fluid nozzle of paint deposits which otherwise build up around the orifice, and the application of heavy paint films at the beginning and end of each stroke is avoided.

For proper triggering, the stroke is started off the work, and the trigger is pulled as the gun approaches the edge of the work. The trigger is released at the opposite edge of the painted area but the stroke is continued for a few inches before reversing direction for the next stroke—see Figs. 5-13 and 5-14.

An experienced painter will accurately hit the edge of the work with the first spray of paint during each stroke, maintaining full coverage over the surface without overspray. However, most painters will rely on a banding technique, as illustrated in Fig. 5-15, to reduce overspray.

The single vertical stroke applied along each end of the panel assures complete coverage and eliminates the waste of paint which results from trying to spray right up to the vertical edge during the horizontal stroke. Along the top and bottom of a panel the gun is aimed at the edge, and thus these automatically are banding strokes. A similar banding operation is employed on edges and outside and inside corners, as illustrated in Fig. 5-15. These operations are always done first, leaving flat surfaces to be sprayed last.

A long panel can be sprayed with vertical strokes, but most sprayers have better control with the more natural horizontal stroke. The panel is sprayed in separate sections 24 to 36 in. long which are overlapped 3 to 4 in. by using the same triggering as with a smaller panel. This is the one operation in which arcing the gun is desired. The experienced painter will feather out each overlap by a short-radius arc at the end of the stroke.

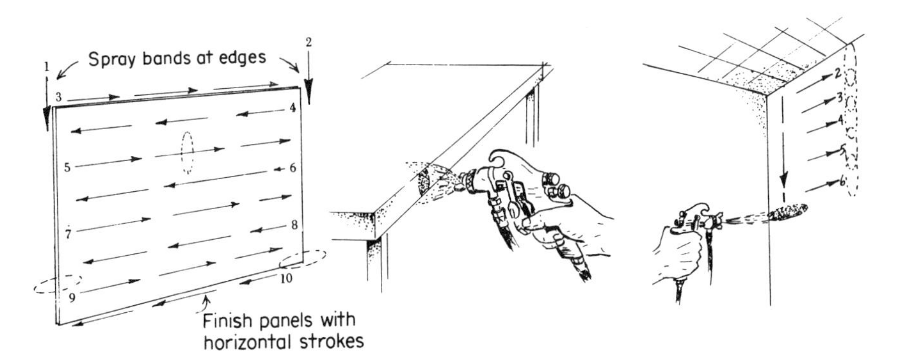

FIG. 5-15 Banding edges and corners.

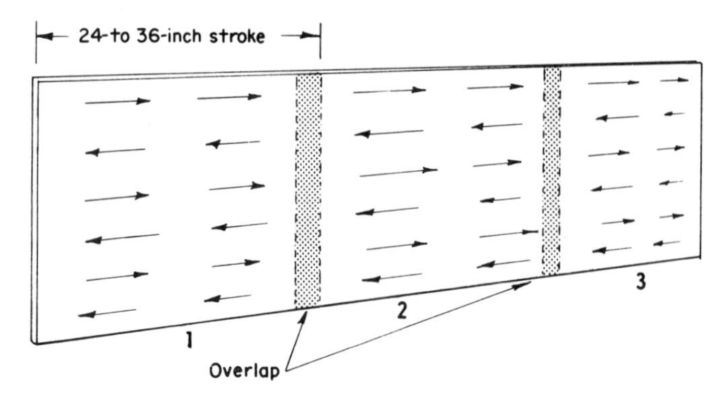

FIG. 5-16 Overlapping strokes.

When spraying a level surface, always start on the near side and work to the far side as shown in Fig. 5-17. This is absolutely essential where lacquer is used since lacquer fog landing on wet work will dry with a sandy appearance. A certain amount of gun tilt is often necessary when painting level surfaces; however, when practical, the work itself should be tilted up so that the spray gun can be held at as nearly a right angle to it as possible.

Faulty results with spray-applied coatings: Even though the application techniques were correct, or reasonably so, the finished paintwork may exhibit faults or failures. A review of the most commonly experienced painting failures, their probable causes and corrective measures, and remedies applied to the paintwork are presented in the following discussion. The illustrations employed in this section were supplied by the E. I. du Pont de Nemours & Co., Wilmington, Delaware, and are used with their permission.

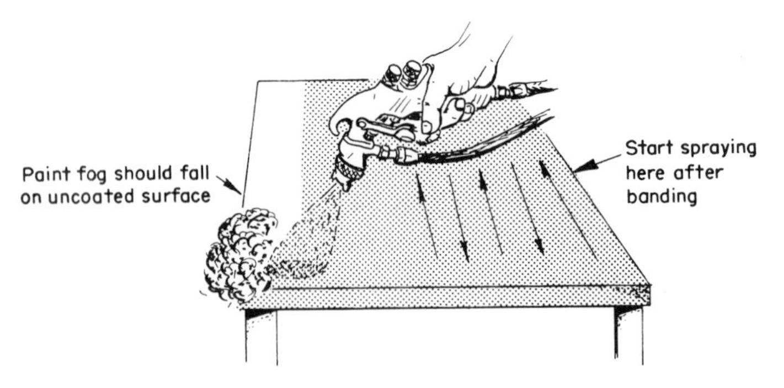

FIG. 5-17 Painting horizontal surfaces.

Orange-peel Appearance

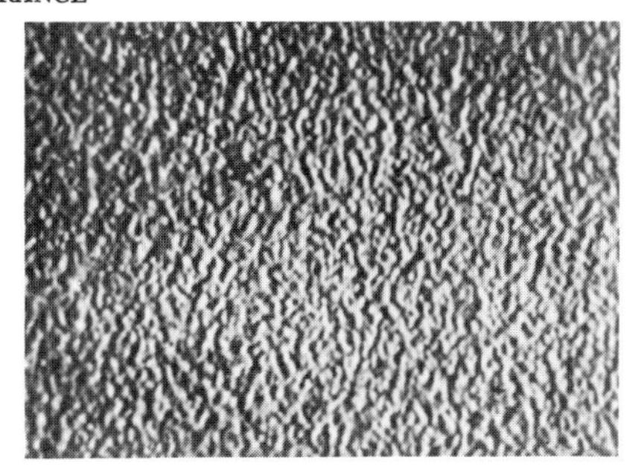

Appearance: Ball peen hammer dents in the surface; paint surface resembles an orange skin

Causes: 1. Improper atomization: atomizing air pressure too low; airless fluid pressures too high; paint too high in viscosity.
2. Sprayed improperly: gun too close to the panel
3. Solvent too fast: evaporates too fast for proper flow-out

Correction: 1. Increase air pressure to the air spray gun; decrease fluid pressure to the airless gun.
2. Follow thinning instructions, using slower solvents.
3. Hold gun at right angle to the work and at a slightly greater distance.

Remedy: Rub with rubbing compound when thoroughly dry; in extreme cases sand down to smooth surface and refinish.

Remarks: In the case of high volume operations with heavy-viscosity coatings applied in thick films per coat, eliminating orange peel becomes increasingly more difficult, if not impossible. The reason for this is that normally there is not enough air available to finely atomize the paint. As the film build increases, the bombardment of succeeding coarsely atomized fluids pockmarks the still soft film previously deposited. However, a moderate orange-peel appearance is acceptable with most maintenance coatings.

Pinholing and Solvent Pops

Appearance: Very fine holes in the paint film, as though caused by pin pricks

Causes: 1. Solvents trapped under the paint surface, breaking through surface to escape
2. Insufficient atomization of fluid.
3. Moisture in atomizing air line

Correction: 1. Do not apply paints too heavily, but spray in uniform, normal coats to allow proper evaporation of solvents.
2. Use recommended thinners.
3. With air atomization spraying, increase air pressure for proper atomization, and service the air line extractor to insure dry air.

Remedy: Sand down to smooth surface and refinish.

FISH EYES—POOR WETTING

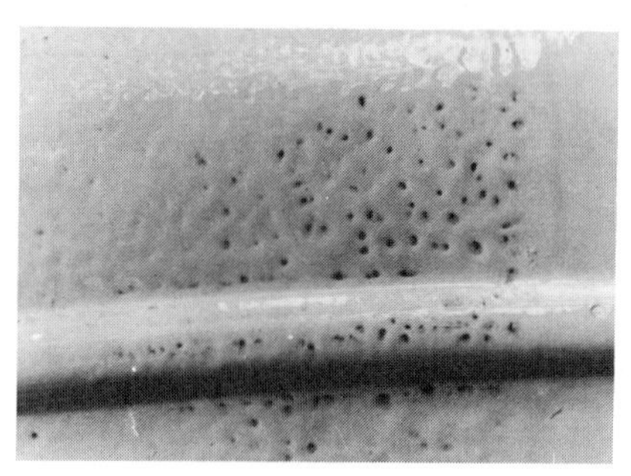

Appearance: Small crater-like holes (fish eyes) or areas of film drawback, exposing undersurface

Cause: Silicone contamination from lubricants, greases, polishes or waxes on the surface being painted.

Correction: Eliminate silicone contamination by washing surface with solvent; add fish eye preventers to the paint being applied.

Remedy: Remove finish in affected area while still wet, and refinish surface after solvent wiping.

RUNS

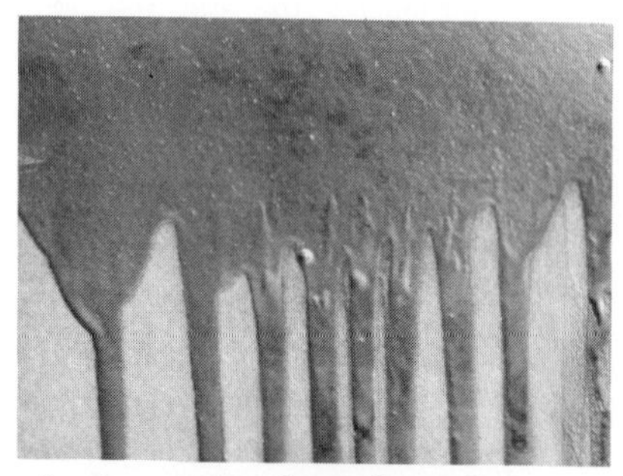

Appearance: Rivulets of paint running down the surface

Causes: 1. Too much or too slow a thinner being used
2. Too wet a coat being applied
3. Cold surface being painted

Correction: 1. Thin paint according to label directions, using less solvents and faster solvents.
2. Regulate fluid adjustment on air spray gun (turn to right to reduce flow of paint); use a smaller airless spray nozzle.
3. Hold the gun at a greater distance from the surface.

Remedy: Sand down to smooth surface and refinish.

Sags

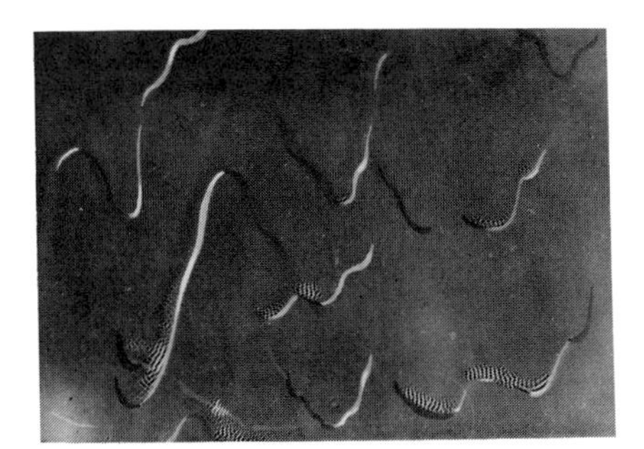

Appearance: Partial slipping of paint film in the form of curtains

Causes:
1. Poor atomization: with air spray application, because of low air pressure or high fluid pressure; with airless spray, due to too low fluid pressure
2. Too heavy a coat applied because gun was held too close to surface
3. Too little thinner used
4. Insufficient drying time permitted between coats

Correction:
1. Adjust air and fluid pressures for proper air spray atomization; increase fluid pressure and perhaps use a smaller nozzle, in airless spray
2. Use proper type and amounts of thinner
3. Hold gun at a greater distance from surface

Remedy: Sand down to smooth surface and refinish.

Blistering

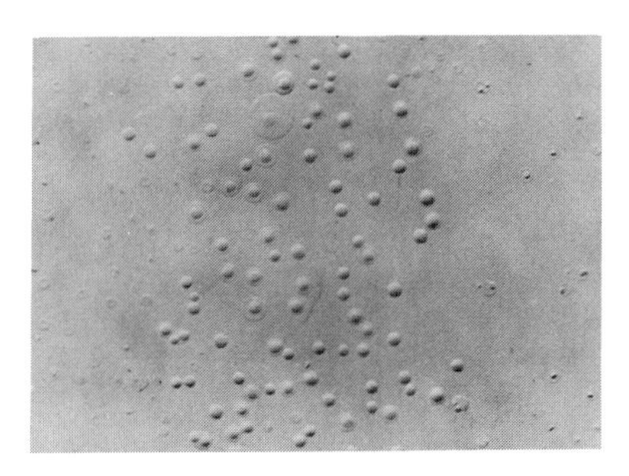

Appearance: Small swelled bubbles or blisters in the paint film

Causes:
1. Rust, oil, or grease on the surface; finger prints on metal
2. Moisture on or in the surface
3. Solvents trapped under a surface skin of dried paint

Correction:
1. Properly degrease surface before painting; keep hands off the surface to be painted
2. Surface must be dry—surface temperature must be above the dew point—before painting
3. Use a slower solvent during hot weather

Remedy: Sand with fine paper to remove blisters in affected areas, or in extreme cases remove finish down to the metal and refinish

MOTTLING: (Appearance similar to Blushing—see illustration below)

Appearance: A nonuniform color, or a mottled metallic appearance in the finish

Causes:
1. Paint not thoroughly mixed
2. Very heavy wet coats applied
3. Electrostatic attraction causing the puddling or lumping of metallic pigments

Correction:
1. Use faster thinners and mix paint thoroughly
2. Reduce fluid pressure or use a smaller fluid nozzle in order to apply lighter coats
3. Hold gun farther from the surface
4. If electrostatic attraction is suspected, use an air hose having a grounding wire to ground the gun

Remedy: Apply an additional topcoat of properly prepared paint

BLUSHING

Appearance: The finish turns "milky" or a lighter shade of color

Causes:
1. Condensation of water in or on the film, which has been cooled through the evaporation of a fast-drying thinner
2. Using a type of thinner incompatible with the paint vehicle

Correction:
1. Add the proper type of "retarder," which is a high boiling, slow evaporating solvent. (Note: during excessive humidity spraying may have to be stopped.)

Remedy: Add "retarder" to the lacquer and apply another double coat

OFFSPRAY—DRY SPRAY

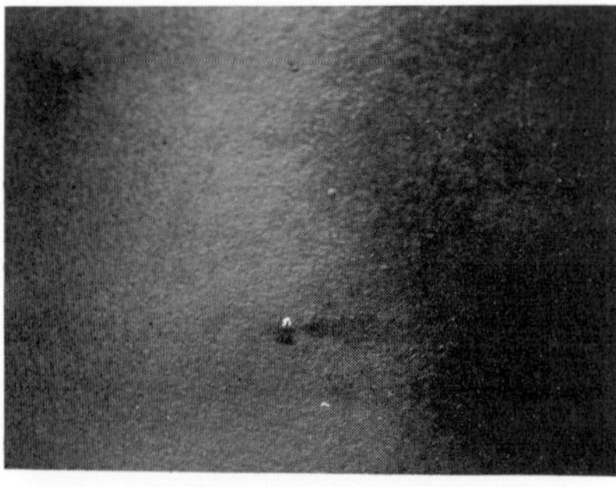

Cause: Dry paint particles falling on the surface outside of the normal spray pattern due to paint fog from rebounding air. This is a problem principally associated with air spraying

Correction: 1. Reduce air pressure
2. Hold gun closer and perpendicular to surface
3. Add slower-evaporating solvents to the paint
4. Dampen the surface before painting with a mist coat of solvent
5. Band edges and spray details first
6. Work in the direction of the air currents in spray booth, and on flat surfaces start on the near side and work to the far side

Remedy: A dry sandy-appearing surface must be sanded down with fine emery paper and recoated

WRINKLING

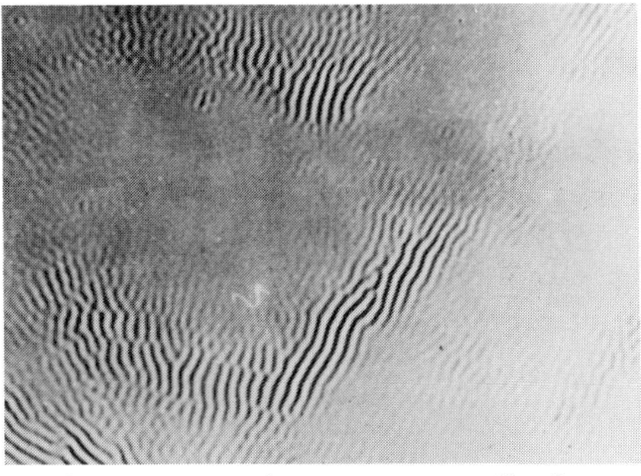

Appearance: A wrinkled surface—one having many small ridges

Causes: 1. Applying very heavy wet coats, particularly of enamels, so that the surface dries and hardens while the body of the film is still soft
2. Applying a paint with strong solvents over an old coating which is softened by these solvents

Correction: 1. Apply thinner films per coat
2. Do not apply finishes using strong solvents over old coatings without testing first for solvent action

Remedy: Remove wrinkled paint and refinish

LIFTING, PEELING

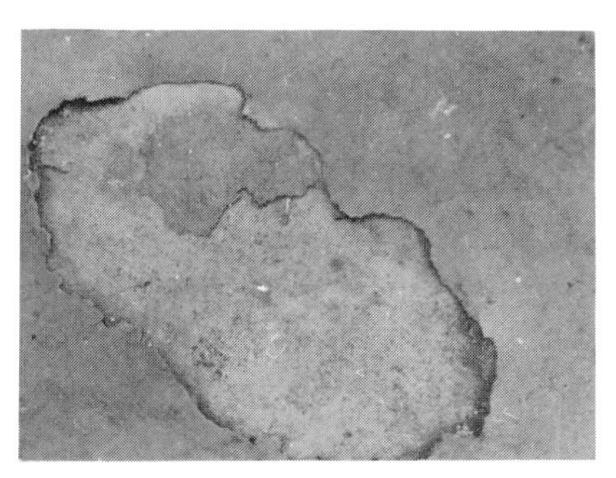

Appearance: Top layer of paint lifting and curling at the edges of cracks

Causes: 1. Improper surface cleaning—oil on the surface
2. Too thorough drying of previously applied protective coat
3. Sandwiching incompatible paint films or primers and topcoats, such as an enamel applied between coats of lacquer

Correction: 1. Correctly prepare surface for paint application
2. Apply second coat within the dry time specified, so that solvents will "bite" into the first coat
3. Apply only compatible types of primers and top coats

Remedy: Completely remove coatings and repaint

Electrostatic spray application: Hand spraying of paint by electrostatic methods will be discussed in this section. Hand guns are available in which the electrostatic charge is applied by either of two processes:

1. The paint is electrically atomized and charged as it leaves the edge of a spinning bell, or
2. The electrical charge is applied to paint particles already atomized either by air spray or airless methods, or applied to the paint stream just prior to atomization.

There are several important advantages in the application of paint by electrostatic methods. These are:

1. Savings of material being applied: Paint loss is greatly reduced through the electrostatic attraction of paint particles to the work surfaces. This attraction is so strong that even the edges and back sides of tubes and rods are coated—an effect termed "wraparound." Paint losses are reduced by 50 to 75 percent over the losses experienced with conventional air spray and airless application methods.
2. Greatly reduced overspray and paint fog: The need for shielding or protecting nearby objects is simplified and sometimes eliminated. Cleanup and equipment maintenance time are reduced.
3. Improved application: With many types of paints, a better quality finish is obtained through the action of electrostatic attraction. Full coating thickness is applied to outside edges and corners.

Most solvent-base free-flowing paints may be applied by the spinning bell method. The flow of coating to the bell is provided by the usual regulated paint supply system. An electric motor in the gun rotates the atomizing bell, with the result that the material flows uniformly to its outer edge.

Atomization occurs under the influence of the electrostatic field as the paint flows from the edge of the bell, forming a spray pattern of electrically charged particles. These charged particles move at slow speeds and tend to deposit on the object at points of maximum electrostatic attraction. The paint application thus will be thinner in cavities and depressions on the surface and heavy on edges or protruding points. This represents a definite advantage when applying protective coatings. Nonelectrostatic methods of paint application (brush or spray) tend to leave thin coatings on edges, where early coating failure usually occurs.

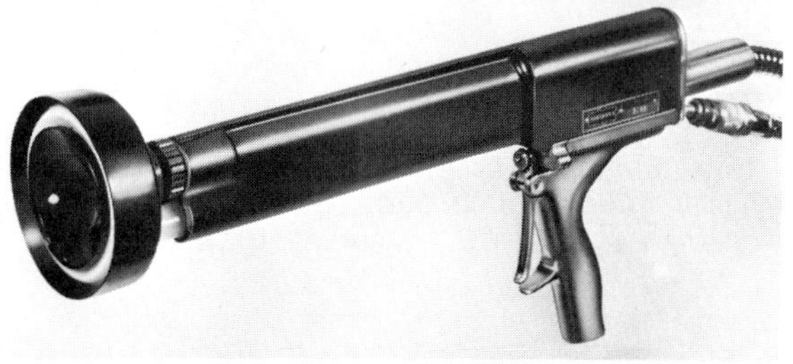

FIG. 5-18 Electrostatic spray outfit—spinning bell type gun. (*Ransburg Electro-Coating Corp., Indianapolis, Indiana.*)

Paint application rates by the spinning bell method are too limited for most high-production-rate applications of maintenance paints and protective coatings. The bell with largest capacity, a 6-in. bell, has an approximate maximum paint delivery rate of 6 oz per min. However, these rates are ample for painting open grills, chain link fences, etc.

Air atomizing and airless electrostatic spray methods are more suitable for the field application of maintenance paints and coatings, since higher deliveries are possible. The forces of electrostatic attraction result in a more uniform coverage over irregular surfaces than is otherwise obtained, and furthermore, these guns may be used as ordinary spray guns whenever desired. These electrostatic spray methods exhibit all the inherent advantages and disadvantages of the conventional air and airless spray methods reviewed earlier in this chapter, to which are added the advantages of electrostatic attraction already enumerated: paint savings; little if any fog and overspray; and less cleanup and protection of nearby objects.

The electrostatic charge is applied either by means of a pointed electrode protruding in front of the gun and extending into the paint atomizing zone, or by an electrode extending into the paint stream in the gun just before atomization. Depending upon the manufacturer, ionizing voltages range from 30,000 to 90,000, with approximately 60,000 volts being employed by several prominent manufacturers. These usually are fixed voltages and cannot be varied. Air spray and airless electrostatic guns may be fed by conventional paint pumps or pressure tanks. However, special electrostatically conductive hose must be employed.

When electrostatic spray equipment is operating properly and the paint conductivity is within the range suitable for electrostatic spraying, a marked wrap around effect is present. If there is only a weak wrap around or none at all the paint conductivity may be too high, the power supply may not be operating, or there may be some electrical trouble with the high voltage cable or gun.

A simple procedure to check for presence of high voltage at the gun electrode in models where the electrode protrudes in front of the gun is as follows:

With the power supply on and with no paint being fed to the gun, hold the trigger depressed and bring the electrode wire at the front

FIG. 5-19 Electrostatic spray gun—air atomizing type. (*Binks Manufacturing Co., Chicago, Illinois.*)

of the nozzle within ½ in. of a grounded object. Watch carefully for a blue corona discharge between the gun electrode and the target, which indicates the presence of high voltage. If no corona discharge can be observed, high voltage is not being delivered to the electrode, and the manufacturers' instructions for checking the electrical system should be followed to determine the fault.

The standard formulations of most solvent-base paints can be sprayed successfully by air and airless electrostatic spray methods, provided that the paint conductivity is not so high as to form a grounding path for the high voltage. Paints are conductive either because conductive (polar) solvents are used or metallic pigments are present. Representative nonpolar and polar (conductive) solvents are presented in Table 5-9.

Paints that are too conductive for air or airless electrostatic spray application seldom can be successfully modified in the field by adding nonconductive (nonpolar) solvents. The paint formulation must be revised by the supplier. All water-base paints and many paints with metallic pigments cannot be applied by electrostatic hand guns because of their high conductivity. (In industrial equipment, water-base paints

TABLE 5-9 Nonpolar and Polar Solvents

I. *Nonpolar and Low-Polarity Solvents (Nonconductive)*

Aliphatic solvents:	Chlorinated solvents:
Mineral spirits	Trichlorethylene
VM and P naphtha	Methylene chloride
	Perchlorethylene
Aromatic solvents:	
Toluene	*n*-Butyl acetate
Xylene	
Coal-tar solvents	Turpentine
Lacquer thinner (aromatic thinner)	

II. *Polar Solvents (Conductive)—Listed in Order of Increasing Polarity*

Methyl ethyl ketone—MEK
Methyl isobutyl ketone—MIBK
Acetone
Isopropyl alcohol
Ethylene glycol ethyl ether
Ethylene glycol methyl ether
Nitropropane
Diacetone alcohol
Ethyl alcohol
Methyl acetate
Water (with usual salt content)

Note: Refer to Table 3-4, page 70, for trade names of the glycol ether solvents.

can be applied electrostatically by insulating the entire paint system from the ground and putting it under the high electrostatic voltage.)

Although air and airless electrostatic hand guns usually can apply paints with standard formulations, optimum results are obtained with formulations having a very low electrical conductivity. Painters say "The deader the better." The use of a conductivity meter for the precise measurement of paint conductivity is recommended for large-scale users of electrostatic spray equipment. They are available from the equipment manufacturer and usually are dimensionally matched to their electrostatic gun.

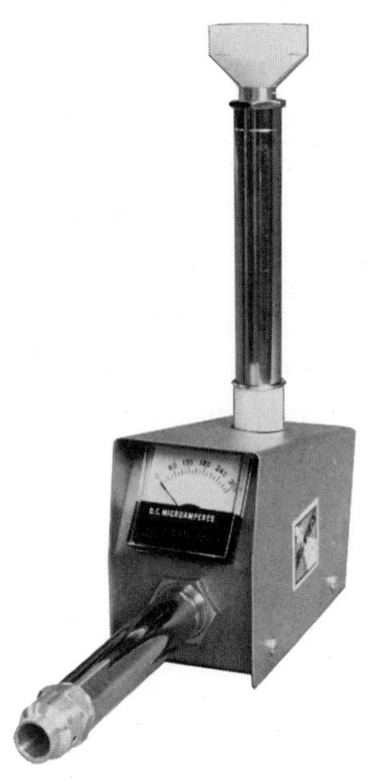

FIG. 5-20 Paint conductivity meter.

Some fairly conductive coatings may be applied successfully by air and airless electrostatic spray hand guns under proper conditions. Metallics and other industrial coatings normally considered too conductive often may be sprayed by formulating with less polar solvents, by use of circulating paint systems, and by other techniques. The gun manufacturer should be consulted for recommendations on the application of specific conductive coatings.

The spinning bell type of electrostatic hand gun atomizes paint electrostatically without air or hydraulic pressures. Best results are obtained with this gun when formulations exhibit a small degree of conductivity. The conductivity of paints to be applied by this type of electrostatic gun often can be adjusted satisfactorily by adding compatible polar or nonpolar solvents. The equipment manufacturers' recommendations should be followed in making this adjustment.

In addition to considerations of paint conductivity, it has been found that optimum electrostatic attraction results when the paint viscosity is adjusted to from 20 to 24 sec No. 2 Zahn cup (14 to 22 sec No. 4 Ford cup). This is lighter than the viscosity of most paints supplied for conventional spraying. Xylene or other nonpolar or low polar solvents compatible with the paint system are recommended for thinning purposes.

Since lower spray pressures generally are used, the charged paint particles move more slowly to the surface being painted than when applied by conventional methods. Consequently, slower drying solvents must be used in the paint formulation to assure that a wet paint film is being applied. Furthermore, a paint particle which dries enroute to the surface quickly loses its charge, and so will not be electrostatically attracted to the surface. Evaporation rates of solvents are presented in Table 5-2 on page 126.

Operation of Electrostatic Equipment. Although high voltages are used, the equipment when properly set up is safe to operate. There is no voltage applied to the gun electrode when the trigger is not pulled. Upon pulling the trigger high voltage is present, but with most equipment this decreases as the electrode approaches a grounded object until, when contact is made, the voltage difference becomes zero. Consequently, holding the gun in one hand and touching the electrode with the other will produce no shock.

The fluid hose must contain a special grounding wire, or the jacket must be conductive to electrostatic charges. In air-atomized spray, the fluid hose usually is constructed so that the jacket, or outer covering, is electrostatically conductive. In airless spray, the hose contains a conductive ground wire embedded in the carcass.

The painter is grounded when gripping the gun handle, since the handle is connected to ground through the high voltage cable, as well as through the hose. Consequently, the painter cannot wear gloves which would insulate his hand from the gun handle unless he first cuts out the palm of the glove.

A power supply unit provides high voltage (low amperage) electrical potential to the gun through a high voltage cable. In some units, the atomizing air supply for electrostatic guns passes through the power supply unit through an air-operated electrical switch. With equipment of this type, high voltage is applied to the gun electrode only when triggering the gun releases a flow of atomizing air.

The object to be sprayed must be grounded, and the surface must be electrically conductive. Wooden structures can be sprayed successfully if the moisture content is around 15 percent or more. A special pretreatment may be applied before painting when it is desired to electrostatically paint dry wood, glass, plastics, rubber, and other nonconductive objects.

Application techniques are simpler with electrostatic spraying than with straight air or airless spraying. Lapping is less critical in applying an even coat and, for many applications, careful attention to triggering is not necessary. Overspray is essentially eliminated through the forces of electrostatic attraction.

With air electrostatic spraying, lower atomizing air pressures are required, both because of thinning and because the electrostatic charge aids in paint particle formation. Likewise, considerably lower fluid pressures are necessary with airless electrostatic spraying, and the degree of atomization usually is finer than without the electrostatic charge.

Because the viscosity of most paints to be applied electrostatically has been adjusted to the desired range, only two or three air spray nozzle setups satisfy most air electrostatic application requirements. A selection of airless spray nozzles is needed, however, depending upon the object shape and application rates desired.

Recommended airless nozzles for various electrostatic spraying applications are presented in Table 5-10.

The inclusion of a preorifice spray insert is helpful with paints that are difficult to atomize to a fine spray without using excessive fluid pressure. The insert tends to reduce the forward velocity of the paint through the nozzle, providing a "softer" spray with increased electrostatic efficiency. An insert may be useful for spraying tubular and open ware. The spray insert orifice size should be equal to or slightly larger than the nozzle size—never smaller.

When painting, it is recommended to hold the electrostatic gun at the usual distance from the surface: 6 to 10 in. with air spraying and when using the spinning bell gun, and 8 to 12 in. for airless atomization. The distance never should be much greater than 12 in., however, otherwise the sprayed paint particle may be attracted preferentially to the

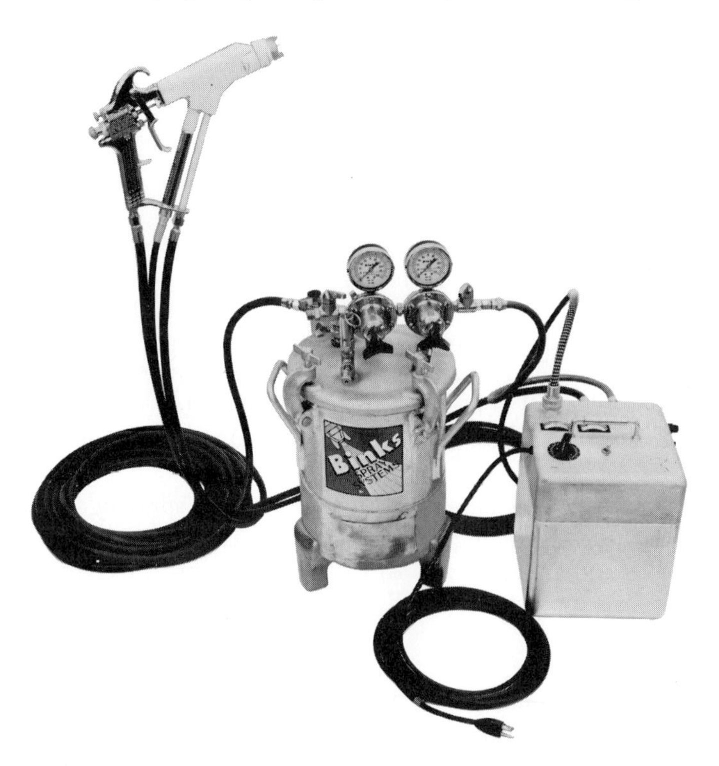

FIG. 5-21 Electrostatic spray outfit—air spray type gun. (*Binks Manufacturing Co., Chicago, Illinois.*)

TABLE 5-10 Airless Nozzle Selection Guide for Electrostatic Spraying

Large targets	Small targets	Tubular and open ware
Recommended fan angle		
65–80° A larger fan reduces the number of passes required.	25–40° Select a smaller fan to reduce overspray that occurs when fan is wider than parts—smaller fans penetrate recesses better.	15–25° A fan should be 3 to 5 times the part width—for maximum wrap-around, about ⅔ of paint should miss front side of the part.
Recommended nozzle size		
Low production: (10–15 sq ft/min) Try 0.011-in. orifice.	(8–12 sq ft/min) Try 0.009-in. orifice.	(5–9 sq ft/min) Try 0.007-in. orifice.
Medium production: (15–20 sq ft/min) Try 0.013-in. orifice.	(12–16 sq ft/min) Try 0.011-in. orifice.	(9–15 sq ft/min) Try 0.009-in. orifice.
High production: (20 sq ft and over/min) Try 0.015-in. orifice. For top volume with some drop in quality, try 0.018-in. orifice.	(16–21 sq ft/min) Try 0.013-in. orifice.	(15–21 sq ft/min) Try 0.011-in. orifice.

painter's grounded hand rather than to the object being painted. This is one reason for the somewhat longer gun barrel on air and airless electrostatic spray guns, since with shorter gun barrels "wrap back" to the painter's hand would be annoyingly frequent.

Paint must be kept from accumulating on the electrode wire and on the face of the nozzle. Paint deposits may be removed with a bristle brush and solvent, *but only after the high voltage has been turned off*. For more extensive maintenance and service of electrostatic spray equipment, refer to the manufacturers' service manuals.

Safety. There is very little paint fire hazard with modern electrostatic spray equipment that has been properly installed and is operated as recommended. Although gun voltages are high, only microamperes of current are required in charging the paint particles. It has been demonstrated that there is so little energy present in a spark from the gun electrode of some equipment that tests have failed to ignite hexane vapors.

However, the control of energy in the electrostatic equipment does not apply to nearby conductive objects that may be insulated from ground. These objects can develop high voltages with appreciable energy as they are contacted by charged air molecules and paint particles. A spark from such objects easily may contain sufficient energy to ignite solvent vapors. Consequently, *it is essential* that all electrically conductive objects—including personnel—be grounded when located within 10 to 15 ft of the gun operating area.

The painter is grounded by his bare-handed grip of the gun; helpers must be grounded by wearing shoes with leather soles. Shoes with insulating soles of rubber, composition, and cork never should be worn. Floors must be conductive; a grounded steel grid is recommended over dry wood floors.

Metallic items must be removed from pockets, such as coins, keys, pencils, nail clips, etc. There have been occasions where these overlooked items have acquired a sufficient charge to ignite solvent vapors as they sparked through the pocket when the painter moved close to a grounded paint bucket.

All metal objects—especially paint and solvent containers—*must be grounded by attaching ground wires.* In addition to grounding the power supply, it is recommended that this unit be located as far as possible from the spray area—at least 20 ft distant.

In addition to the careful attention to grounding stressed above, all other safety precautions should be observed which normally apply to air or airless spray application of paint. Refer to earlier discussions in this chapter and to Chap. 3 for details. For additional safety recommendations, also refer to *NFPA* Bulletin No. 33 "Spray Finishing," issued by the National Fire Protection Association, Boston, Massachusetts.

Spray application of multicomponent coatings: For those catalyzed coatings which have the very short pot life of only seconds or minutes, special equipment has been developed which permits their application at full production rates. These quick-setting coatings—polyesters, polyurethanes, and certain epoxies—exhibit such a short life before gelling that the addition of catalyst, mixing, and application all must be accomplished at the spray gun.

Several application systems have been developed for applying polyester coatings—either "gel coats" having no reinforcement, or "FRP" layups with fiber glass reinforcement. When fiber glass reinforcement is desired, a fiberglass chopper is attached to the spray gun which feeds chopped glass fibers into the resin spray stream.

The three most commonly employed methods for the spray application of polyester resins are:

1. Simultaneously spraying two polyester resin solutions by means of a two-headed spray gun. The one spray consists of a styrene solution of resin with promoter, and the other spray of resin solution containing the MEK peroxide catalyst. The spray patterns converge

and mix before reaching the surface. These solutions are available from coating formulators already prepared for this method of application.

Usually, pressure pots are employed to feed a double-headed air spray gun; coupled high pressure pumps are required for airless spray application. These pumps are linked together on a crossarm arrangement to assure a constant ratio between their pumping rates, and to permit adjustment of the pumping rates without changing this ratio.

2. Injecting a solution of the MEK peroxide catalyst at a carefully controlled rate into the atomized zone of sprayed resin/promoter solution by means of a catalyst side-injection gun. The resin can be atomized either by air spray or by airless methods. The catalyst injector is in effect an attachment to a standard air spray or airless gun.

The injected catalyst consists of a 33 percent solution of MEK peroxide in a suitable ester such as ethyl acetate. A stainless steel pressure tank is employed for the catalyst solution and tanks or pumps are used to feed the polyester resin/promoter solution.

3. Air spraying the resin/promoter solution, with the atomizing air carrying the volatilized MEK peroxide catalyst. The catalyst solution is evaporated under controlled conditions as the atomizing air passes through the stainless steel catalyst container, called a "catalyzer."

Safety. In all application systems for polyesters, it is *essential* that only stainless steel (or aluminum) metal surfaces be allowed to contact the MEK peroxide solutions. This peroxide may decompose violently if contacted by brass or steel parts. All gun nozzles and fittings, hose connections, pressure tanks, pumps, and all other metal parts contacting the peroxide solution are stainless steel, or in some cases aluminum. In

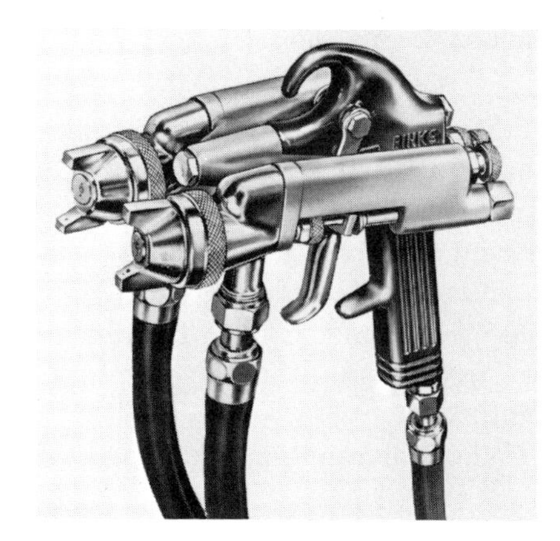

Fig. 5-22 Two-headed air spray gun for polyester application. (*Binks Manufacturing Co., Chicago, Illinois.*)

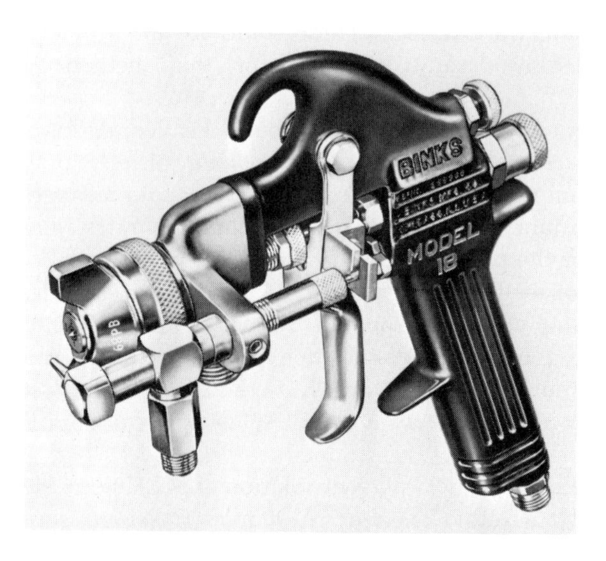

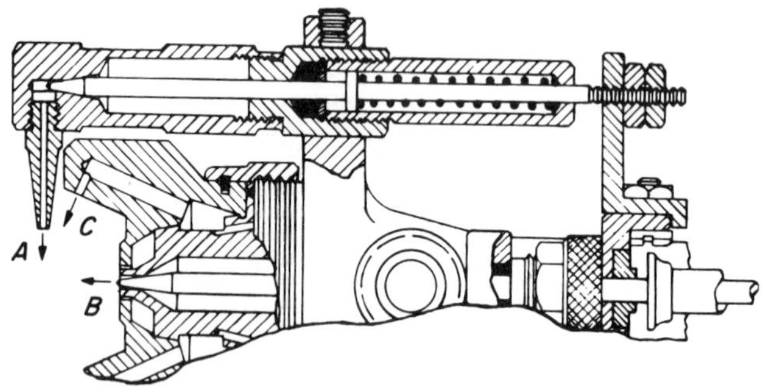

FIG. 5-23 Catalyst side injection gun for polyester application. The mixture of MEK peroxide and ethyl acetate is fed under pressure from catalyst nozzle A, where it meets the resin stream coming from fluid nozzle B. At the junction point, the two streams are atomized and mixed by air from air nozzle C. (*Binks Manufacturing Co., Chicago, Illinois.*)

addition, the MEK peroxide solution also may become unstable if pressured to airless spray pressures—consequently in all application methods reviewed above, airless spraying applies only to the resin/promoter component of the coating.

Since the volatile styrene monomer is toxic, the painter and helpers must be protected from its inhalation through use of air-supplied hoods or respirators. Cartridge-type respirators do not afford sufficient protection.

Where more intimate mixing of the catalyst with resin base is required, such as with the polyurethanes and certain epoxy formulations, the systems reviewed above for polyesters will not be satisfactory. Two-package polyurethane coatings and foams may be mixed and applied by both air spray and airless methods by means of the following equipment:

1. Air spray application with an internal mixing gun: The two solutions, polyisocyanate and polyol, are fed separately to the two gun fluid inlets by means of a low-pressure double-linked pump (similar to the high pressure two-component pump described for polyesters, above).
The two streams are blended in a mixing chamber at the front of the gun through the action of a bladed rotor arm rotating at 20,000 rpm. The mixed material is then ejected through the spray nozzle located at the far end of the mixing chamber. Both internal mix and external mix spray heads are available.
The mixing head includes provisions for air or Freon feed for urethane foam production. Also, a solvent-feed connection is provided since it is very important to solvent-flush the fluid passages of this gun immediately after each spray of resin.

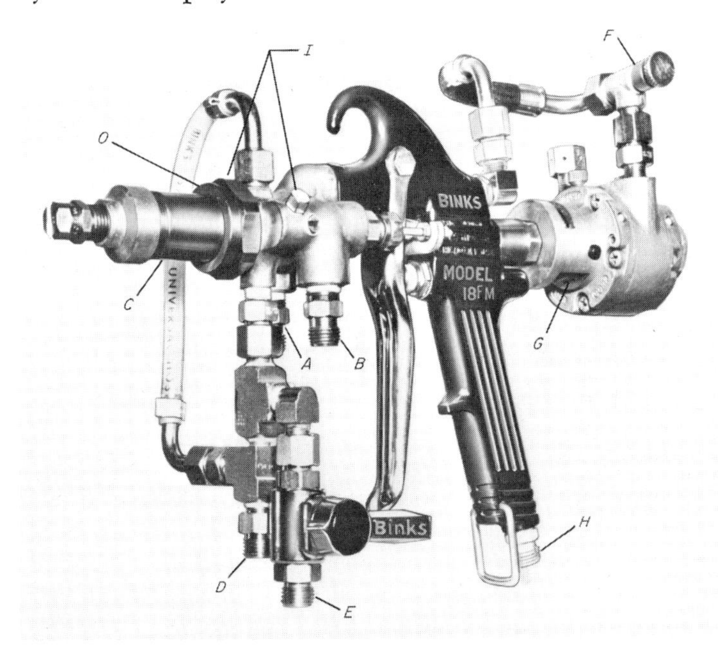

FIG. 5-24 **Internal mixing spray gun for polyurethane application. (*A*) Isocyanate resin inlet; (*B*) polyol inlet; (*C*) mixer chamber; (*D*) spray air inlet (adjusting valve optional); (*E*) solvent flush inlet with valve; (*F*) air motor adjusting valve; (*G*) manual air motor starter wheel; (*H*) air motor operating air inlet; (*I*) solvent—air flush check valves (flushes both resin ports).** (*Binks Manufacturing Co., Chicago, Illinois.*)

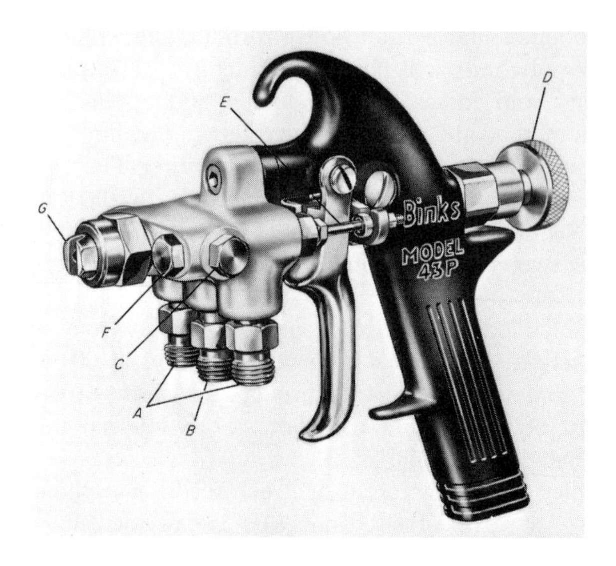

FIG. 5-25 Double-feed airless spray gun for both polyester and polyurethane application. (*A*) **Connections for A and B resin solutions;** (*B*) **solvent flush inlet connection;** (*C*) **solution impingement orifices (removable);** (*D*) **solvent flush control knob;** (*E*) **fluid valve stems (two);** (*F*) **gun head cleanout ports (two);** (*G*) **standard airless spray nozzle.** (*Binks Manufacturing Co., Chicago, Illinois.*)

2. Airless application of polyurethanes with a two-feed airless gun: The polyisocyanate and polyol solutions are pumped separately to the two fluid inlets in the head of this airless spray gun. Here they meet as high velocity impinging streams, mixing thoroughly in the resultant turbulence before being ejected through the spray nozzle. A third inlet for solvent flushing also is provided.

Brush Application of Paint

Although brushing of paints would seem not to need any explanation, there is a difference between the work of an experienced painter and that of an amateur. This section presents some suggestions with respect to the selection and care of brushes, and reviews the techniques of brush application as acquired by the skilled craftsman.

It generally is considered good practice to apply the first primer coat of any protective coating system by brush in order to obtain a more intimate contact of the first coat with the surface. Brushing works the primer into pits, crevices, and pores, and also works any suface dust into the body of the paint. Brush application is preferred for applying paint to small areas and trim, for most interior work, and for all paint application where the advantages of spraying would be lost by the need for masking of surfaces not to be painted.

Equipment: The best paint and varnish brushes generally are made from hog bristles. The outer ends of the bristles are split into two or more fine branches, which results in finer brush marks and greater paint holding ability. Fine lettering and artist's brushes may be of squirrel hair (so-called camel's hair). Whitewash brushes are constructed of mixed hair and tampico fibers. Nylon brushes have a cost advantage over hog bristle brushes and are far more wear resistant, which is particularly evident when painting abrasive surfaces. Nylon bristles have been developed to the point where they may be used to apply all types of paint.

The size and shape of the brush are chiefly a matter of the personal preference of the painter. Flat brushes are preferred for expedient work on flat surfaces and are by far the most commonly employed brush. Bristles sometimes are shaped for specific applications, such as the use of tapered bristles on sash brushes. Round or oval brushes are often preferred for rough surfaces, for painting rivet heads, and for constricted areas.

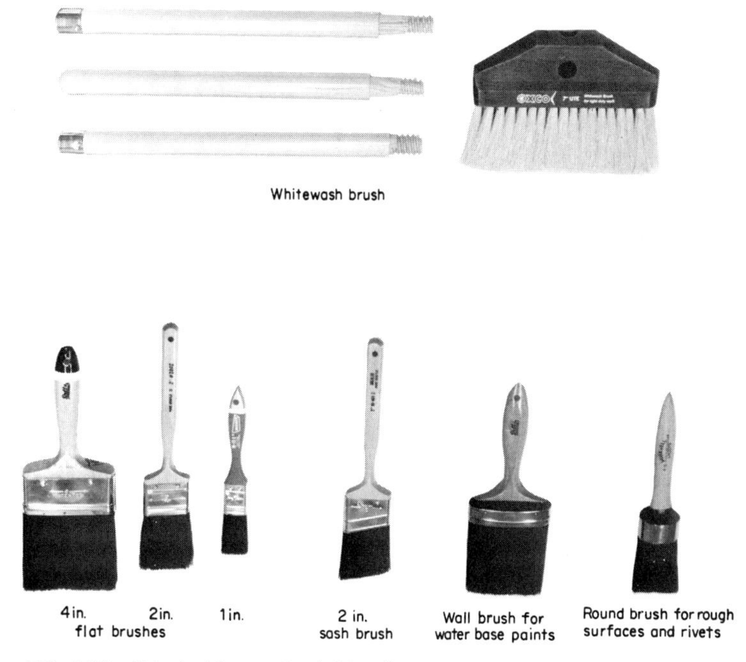

FIG. 5-26 Principal types of paint brushes.

The size of the brush must, of course, be adapted to the work being painted. A 4-in. flat brush is considered the maximum width for effective painting, except Kalsomine and whitewash brushes, which may be 6 or 8 in. wide.

New high quality brushes should be conditioned by suspending the bristles in raw linseed oil for one or two days before use. The brush should then be cleaned in mineral spirits or turpentine before use.

For brushes to serve their full life, it is essential that they be properly cleaned at the end of every day's use. Wash the brushes in thinner and work surplus thinner and pigment out of the brush by placing the brush against a flat surface and bending the bristles back at the heel. This should be repeated until all of the paint is removed from the heel of the brush. The brush is then combed out and dried with a clean cloth. Never allow brushes to stand on the bristles in a container of paint or solvent.

Brushes used for applying water-base paints shall be thoroughly washed with water and detergent immediately after use and allowed to dry. Metal-bound brushes should never be suspended in water, as water is apt to burst the binding.

Application techniques: The brush should not be dipped deep into the paint, as this results in overloading the bristles and filling the heel with paint. The brush is dabbed on the side of the can in order to distribute the paint among the bristles. A good craftsman uses one side of the can for this purpose, and keeps the other side clean. The handles of his brushes, when laid at rest across the can, are placed on the clean side.

The paint should be applied in a uniform film at the specified spreading rate. Do not attempt to cover too large an area at a time, but work in strips about a foot wide and keep the edges wet. Cover the work completely to avoid retouching afterwards. It is desirable to apply successive coats by cross-brushing, which minimizes brush marks.

Hold the brush at an angle of about 60° to the work. Make several light strokes in the area to be painted, then spread the paint crosswise with a moderate, even pressure and finish off crosswise again with longer, lighter strokes. When using a round brush to apply paint to badly pitted steel, corners, rivet heads, and similar surfaces, the initial spreading should be in a pattern of small circles which completely cover the area.

Oleoresinous paints should be brushed out well to completely coat all irregularities on the surface. Lacquer-type rapid-drying paints will not permit as much brushing and cross-lapping. These paints should be spread with a few rapid strokes over a small area, which should then be left alone.

Brush marks will be less conspicuous when the laying-off is done towards the lighting on large areas such as ceilings and walls. On woodwork the laying-off should be in the direction of the grain of the wood. Thick paint coats on edges should be avoided by brushing towards the edges rather than away from them.

Runs, sags, and drips may be avoided by applying no more paint than the surface will support and spreading paint uniformly. When runs or sags do occur, they should be brushed out promptly. Such defects are not only unsightly, but also may give rise to other troubles such as improper drying, wrinkling, and porosity.

The rate of application by brush varies greatly with the shape, size, and surface condition of the object being painted. On large, flat, smooth surfaces, brush painting 200 sq ft per hr is considered a fair rate.

Roller Application of Paint

This specification covers the use of rollers and stipplers for paint application. Roller application is becoming an increasingly popular alternate to brushing for the application of interior house paints, particularly water-base wall paints. When paint fog would require elaborate protective measures, pressure-feed roller coaters may be employed in place of sprayers for painting exterior surfaces.

Rollers are recommended for applying paint to flat interior surfaces such as walls and ceilings, especially when the surfaces have an uneven texture such as those of brick, plaster, and acoustical tile. Furthermore, the roller method is particularly well-adapted for painting open work such as grills and wire net fencing by using a long nap roller.

The roller coater also is used to create a decorative stippled effect through choice of a wide selection of textured rollers, and narrow rollers are available for special stencilling effects.

Equipment: There are many types of rollers and roller covers on the market designed to meet individual requirements.

The hand roller consists basically of a simple handle with metal roller core and a removable roller cover. The better types are provided with ball bearings. The roller is used with a corrugated or dimpled aluminum or enameled steel tray having a sloping bottom, which holds the paint. After dipping the roller in the paint, the roller is run over the sloping bottom several times to remove excess paint and to distribute the paint evenly over the roller surface.

With the pressure-feed roller coater, paint is fed from a pressure tank through a hose to the roller handle, which connects with the hollow perforated roller core. A valve on the handle controls paint volume.

Roller covers are made of lambs wool, frieze, mohair, or synthetic fabric wound on plastic or fiberboard cylinders. The cover material and the height of the nap govern the texture of the finish. Regular lambs wool with a ⅜-in. nap is used on smooth or semismooth surfaces and gives a very slight orange-peel effect. The 1-in.-nap lambs wool cover is used on stucco surfaces and for fence painting. Hard twist carpet covers are used for stipple finishes, and frieze covers are used for texture paint.

The normal roller is 7 to 9 in. long. Narrow rollers of 2 or 3 in. long may be used for stencilling or trimwork, although their efficiency for the latter purpose is questionable. Special types have been developed to conform to pipe and other curved surfaces.

Application techniques: The hand roller is dipped into the paint tray and rolled over the sloping bottom to squeeze out excess paint. The operator must learn to get the correct amount of paint on the roller and to apply an even coat of uniform thickness by regulating the pressure on the roller while covering the surface. To obtain complete coverage, especially on rough surfaces, the roller must be worked in several directions back and forth. All final strokes should be made systematically in one direction.

Because the normal roller cannot apply paint right up to an obstruction

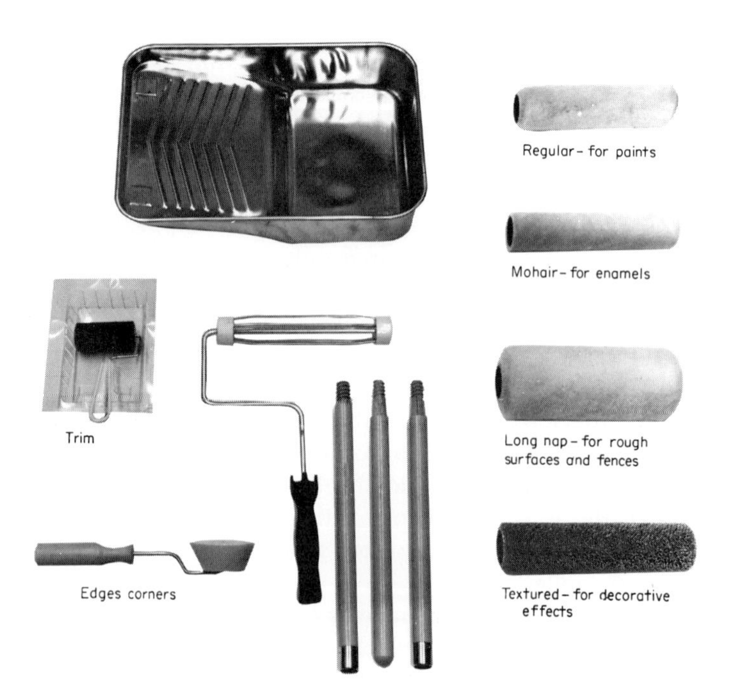

FIG. 5-27 Examples of roller coaters.

or interior corner, the surfaces to be painted first should be "banded." Paint should be applied by brush along all edges and interior corners before roller coating the flat surfaces.

Application of Filled Mastics and Hot Melt Enamels

Heavy-bodied filled coatings, as well as putties and caulking compounds, generally are applied by specially designed high-pressure extrusion and spray equipment. Hot melt enamels usually are hand-applied by mopping, daubing, and troweling. Since many of these materials are employed for heavy-duty corrosion protection, their application will be discussed briefly in this section.

Extrusion and spray application of nonflowing materials: High-pressure pumps have been designed to handle nonflowing and filled materials. These pumps are characterized by large valves and fluid passages so as not to restrict flow, and also to allow passage of the fibrous or aggregate-type fillers.

These materials do not seek their own level as they are pumped from their container, normally a 5-gal pail or 55-gal open head drum. Consequently, "follower plates" are employed, which are rubber-rimmed metal plates which fit in the container and rest on top of the material. Follower plates wipe the container walls clean as the material is withdrawn.

The pump usually is mounted on a framework, which slides up and down and rests on the follower plate. In some models, the pump mount may be lifted, or both lifted and forced downward, by air pistons located in the vertical tubular framework. Pump priming and feed are assured by a "shovel," which is a plate piston loosely fitting in a cone cylinder located under the follower plate and immersed in the material.

Large-diameter high-pressure hose must be used. Wire braid hydraulic hose in ½-in. and ¾-in. sizes is suitable for most mastics and putties, since these usually are oil-base materials. When greater solvent resistance is required, large-diameter high-pressure reinforced nylon hose is used.

A coarse atomization is produced by air spray guns designed with large fluid passages. The spray is sufficiently fine for most filled coatings where spray application is desired—underbody coatings, for example. However, in most applications these materials are flow-coated or extruded, using airless guns having a large selection of interchangeable nozzle shapes to meet the application requirements.

Daubing and troweling of hot melt enamels: Hot-applied coal-tar enamel, hot-applied asphalt, nonskid floor coatings, and other very viscous coatings may be applied by Tampico fiber daubers or by pouring and troweling.

Daubers shall be made of horsehair or high-quality Tampico fibers set in solid hardwood blocks with handles, and shall be of the size best adapted for the work to be coated. Mops made of fiber glass also have been used successfully to spread thick coatings on floor surfaces. Trowels having a rectangular steel blade and sturdy handle are used to spread poured-on coatings over horizontal surfaces.

The application of hot-applied coal-tar enamel or asphalt requires suitable heating kettles with accurate and easily read thermometers. For hand application a small-size heating kettle of not over 50-gal capacity should be used, because the enamel should not be kept at the application temperature for a longer period than necessary. Iron paddles may be used for stirring.

Buckets are filled from the heating kettles with ladles. These buckets are used to pour the enamel on the area being coated. Electrically heated or insulated buckets with a capacity of up to 2¼ gal are recommended for hand daubing.

Daubing, mopping, or troweling are not synonymous with indiscriminate smearing, but in fact require special skill to obtain well-bonded

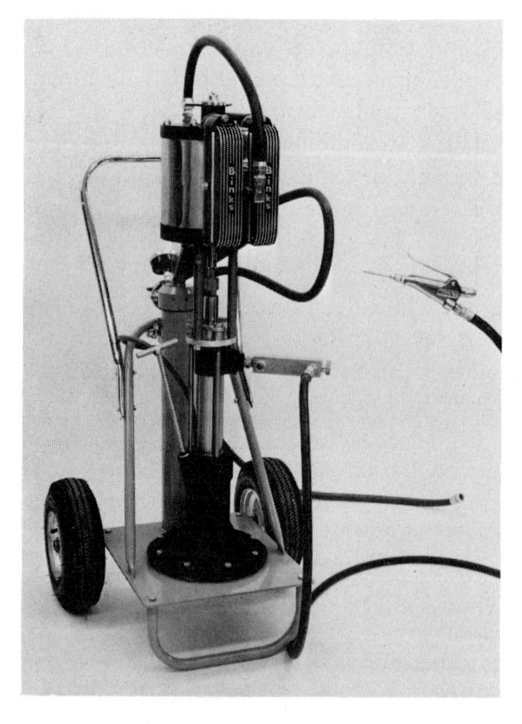

FIG. 5-28 Extrusion application outfit. (*Binks Manufacturing Co., Chicago, Illinois.*)

uniform coatings without excessive holidays or thin spots. The application by daubing of hot coal-tar enamels should be done only by experienced workmen because the application temperature is of great importance. It is not a job for painters or laborers untrained in this operation.

The melting kettle shall be as close as possible to the work so that the temperature drop shall be less than 50°F in the transfer from the kettle to the work. The application shall be done in a systematical manner, covering small rectangular areas with short strokes. Three strokes shall overlap and form a continuous coating. When coating surfaces joined by welds, the strokes of the first coat shall be applied along the weld.

A second coat shall be applied in the same manner with the final strokes at right angles to the first coat. As an alternate, a double coat may be applied by the double lap or "shingling" method. By this method the strokes are laid and overlap as shingles on a roof.

Safety considerations: When handling coal-tar primers and enamels, all exposed skin areas should be protected with a suitable cream. In addition, hands should be protected by leather gloves. Vapors from heated coal-tar enamels and coal-tar solvents are not only toxic but also are strongly irritating to the skin and eyes. Vaporproof goggles should be worn, and respirators are desired.

Because of the temperature of the enamel it is essential that care be taken to avoid burns. In the event of a bad burn, do not try to remove the enamel but go to a doctor as quickly as possible. In case of minor burns by spattered enamel, strip off the enamel and apply a protective cream. Grease will serve as a temporary substitute.

Chapter 6

Painting Economics

Introduction

Paints and coatings are applied for four principal reasons:

1. Decoration (for appearance, and for psychological effects)
2. Identification (for safety, and to identify contents)
3. Protection (protecting the substrate against deterioration)
4. Sanitation (protecting the product against contamination)

and the economic considerations which apply in each case differ considerably.

The discussion of painting economics in this chapter will be concerned primarily with the application of paints and coatings for reasons of decoration and protection, and in particular when applied by professional painters in industrial applications. Even with these limitations, painting economics is a complex subject strongly influenced by local and specific conditions, and so only guidelines can be presented for consideration.

The nonprofessional painter, such as the do-it-yourself home owner, is faced with fewer and simpler economic decisions. Surface preparation will be by hand or rented power-cleaning tools, and paint application almost invariably will be by brush or roller coater. He is supplying his own labor, so his problem usually is simply that of choosing the proper paint system of desired quality. Discussions in Chap. 1 and in this chapter under "Selection of Coating" will assist him in this matter.

The economics of maintenance painting, however, are more involved. Four principal cost factors must be taken into consideration:

1. Selection and cost of the paint or coating
2. Surface preparation costs
3. Paint application costs
4. Maintenance costs

These factors are all interrelated. The proper type of coating first must be selected to provide a satisfactory service life, otherwise recoating is required with uneconomic frequency. However, the best material for the service will be wasted without proper surface preparation and application. A system which can be economically maintained will result only when proper consideration has been given to all three factors: material, preparation, and application.

With maintenance paints and coatings, the economic consideration of greatest importance is that of cost over a number of years, and not simply that of first application cost. Is it cheaper to invest in an expensive high-quality initial application which is expected to show low maintenance costs over a number of years, or to invest in a less costly first application which must be recoated at an early date?

On an average, paint material costs are only ¼ to ⅙ of the total applied cost. It is recommended that for modest-sized jobs the best coating system should be selected and applied under proper conditions, so as to reduce maintenance to a minimum. In this case, direct dollar costs are considered. With industrial painting programs, however, the approach is not so simple. Economic studies which involve considerations of dollar value must be made on the projected costs over a number years.

The value of the dollar is not constant. Its potential earning power, inflation, and taxes all affect the future value of today's dollar. In projecting an industrial maintenance painting program, the future worth of today's dollar must be estimated. In other words, what is the dollar which is planned to be spent 5 or 10 years from now worth today? By not spending a dollar today and investing it instead, $1.70 to $1.80 would be available for spending around 10 years from now, on a future maintenance program.

Considerations of dollar value applied to painting economics indicate that it may be best to limit the size of the initial investment, and instead to spend future maintenance dollars. Such conclusions should be approached with caution under present conditions of escalating labor costs, especially if the dollar value approach leads to what appears to be unreasonable results.

There are several methods by which the present worth of future dollars may be estimated. This is a complex matter beyond the scope of this book. The interested reader is referred to an excellent series of articles on this subject published in 1967 by the National Association of Corrosion Engineers, Houston, Texas, entitled "Corrosion Control Makes Dollars and Sense."

The discussion of painting economics presented in this chapter will include a brief review of each of the four principal factors concerned with painting economics mentioned above: coating selection, surface preparation, application, and the applied coating system and its maintenance.

Selection of the Coating

The premature failure of paints and coatings may be due to one or more of several reasons. The coating system may exhibit

1. Poor resistance to the exposure condition (of special importance for coatings in immersion service)
2. Poor resistance to undercutting corrosion (poor choice or application of primer)
3. Poor adhesion to the substrate, or between coats
4. Early failure due to poor application (thin or porous coats applied)

The first two causes of premature coating failure are concerned primarily with the composition of the coating system (primer and topcoats), while the last two factors usually are the results of improper application and will be discussed later on in this chapter.

Composition: The first consideration in selecting a satisfactory paint or coating is to be sure that the paint composition or type is suitable for the service condition under consideration. The paint or coating type may be selected based on

1. Knowledge of the properties and recommended service conditions of the different types of paint and coatings

2. Recommendations of the supplier or manufacturer
3. Knowledge of exposure tests conducted by technical organizations, the manufacturer, or others
4. Results of panel or patch exposure tests conducted by one's own organization

The serviceability of a *protective coating* will depend primarily upon the composition of the vehicle, or in other words, upon the generic type of the coating. Pigments employed are either inert, or inhibitive in the case of primers. The recommended service conditions of various generic types of paints and coatings are presented in Chap. 1 of this book and in outline form in Table 6-1.

Table 6-1 presents the principal recommended exposures or service conditions for paints and coatings. This information is to be used as a reference guide only. Variations in the formulation of protective coatings within a generic type can greatly affect the serviceability—especially when in immersion service.

The serviceability of a *paint* for atmospheric exposure will depend not only upon the composition of the vehicle, but also to a large extent upon the nature of the pigments. Paint pigments and their effect on paint properties are reviewed in Chap. 1 starting on page 31.

It is difficult if not impossible for the small-volume purchaser of paints to judge the quality by data presented on the can label. In purchasing "trade sales" paints it is best to go by the reputation of the manufacturer, or by past experience, if any. Most labels will give the vehicle composition and percent volume, and some indication of pigmentation type and quantity. Since paint coverage depends upon the nonvolatile solids content, the paint label showing a higher percentage of nonvolatile solids would be a better buy if cost and other factors are equal—which is seldom the case. For a more detailed discussion on trade sales paints, the reader is referred to published literature on the subject such as D. F. Householder's Trade Sales Paints, Chap. 29 of C. R. Martens, ed., "Technology of Paints, Varnishes, and Lacquer," Reinhold Book Corporation, New York, 1968.

The manufacturers of protective coatings are very interested in seeing that their product is properly applied and is employed only for suitable service conditions. They usually are the first ones blamed for premature coating failures. Consequently, these manufacturers have available extensive service data and reliable information as to requirements for surface preparation, application, and recommended services. Their technical personnel always should be consulted before selecting a protective coating for a severe service condition.

Exposure tests: Large-scale appliers of paints and coatings often conduct field exposure tests as an aid in selecting the most serviceable

TABLE 6-1 A Guide to the Serviceability of Paints and Coatings

	Oleoresinous				Solvent-dry							Catalyzed								Other	
	Drying oil	Alkyd	Phenolic varnish*	Epoxy ester	Chlorinated rubber	Vinyl	Vinyl acrylic	Silicone*	Silicone alkyd*	Acrylic lacquer	Bituminous†	Epoxy	Bitumen epoxy	Epoxy-phenolic	Phenolic	Polyurethane (2 package)	Urethane (1 package)	Hypalon	Polyester (FRP)	Zinc silicate*‡	Water base§
Atmospheric exposure:																					
Uncontaminated	E	E	E	E						E	E						E			E	G
Industr. and chem.	G	E	E	E	E	E	E			G	E			E			G	E	E	E	G
Humid and marine	G	G	E	E	E	E	E				E	E	E	E			E	E	E	G	
Marine: spray/splash			E	G	E	E	G				E	E	E	E	E	E	G	E	E	X	
Heat resistant			G					E	E		X									E	
Immersion service:																					
Water: distilled					E	E					E	E	E	E	E			E		X	
Water: potable	N	N	G	N	E	E	G	N	N	N	E	G	X	E	E			E		X	
Water: fresh	N	N	G	N	E	E	G	N	N	N	E	G	E	E	E	E	N	E	G	X	N
Water: salt, brine	N	N	G	N	E	E	G	N	N	N	E	E	E	E	E	E	N	E	G	X	N
Hot water	N	N		N				N	N	N	X			E	E		N	E		X	N
Salts	N	N	G	N	E	E	G	N	N	N	E	E	E	E	E	G	N	E	G	G	N
Acids, inorganic	N	N		N	E	G		N	N	N		G		E	E	X	N	E	E	X	N
Acids, organic	N	N		N	X	E	G	N	N	N		G	E	E	E	E	N		X	X	N
Alkalies	N	N	X	N	G	E		N	N	N	E	E	E	E	X	G	N	E	X	X	N
Crude oils	N	N		N				N	N	N	G	E	E	E	E		N		E	E	N
Refined products	N	N	G	N	E	E	G	N	N	N	G	E	E	E	E		N	X	E	E	N
Solvents: aliphatic	N	N	G	N	G	G	X	N	N	N		E	E	E	E	G	N	X	E	E	N
Solvents: aromatic	N	N		N	X	X	X	N	N	N	X	G	X	E	E		N	X		E	N
Solvents: alcohol	N	N		N	G	E		N	N	N	G	G	E	E	E		N		G	E	N
Solvents: ketones	N	N		N	X	X		N	N	N		G	G	G	E		N	X	X	G	N
Solvents: chlorin	N	N		N	X	X		N	N	N			G	G		X	N	G	X	G	N

KEY: E = Generally excellent; G = Good to excellent depending upon specific condition;
X = Unsatisfactory–coating will show short service life. N = Not serviceable.
No rating: Coating generally not used in this service.

Notes:

* Heat-resistant Ratings: Silicones and zinc silicates are rated for atmospheric service above 500°F; silicone-alkyds and others are rated for atmospheric services below 500°F.

† Bituminous Coatings: Atmospheric exposure refers to asphalt and to coal tar with a coal-tar dispersion topcoat; immersion service refers to coal tar only—asphalt is not recommended.

‡ Zinc silicates are rated for service without topcoats. (In most services, these are topcoated with suitable coatings.)

§ Water-base paints are rated for application to steel surfaces.

paint or coating for the service condition in question. Panel and patch testing of protective coatings, when properly conducted, will provide a reliable screening between serviceable and nonserviceable coatings.

Panel testing is a long-range program. The KTA panel developed by the Kenneth Tator Associates, Coraopolis, Pennsylvania, shown in Fig. 6-1, or a modification of this panel, generally is used. This panel provides opportunities for most types of coating failure: undercutting corrosion at the scratch, and failure along edges, welds, and in pockets where dirt and water collect. General overall resistance to the service condition is indicated by the coating on the flat surfaces.

In an extensive paint panel exposure test supervised by the author, modified KTA panels, fabricated of low carbon steel, were employed. Six panels were exposed for each coating system under each exposure condition—three were prepared by the coating manufacturer, and three were prepared by the plant painters. These were supported in redwood frames for immersion service, and by insulators on metal racks inclined at a 45° angle for atmospheric exposures. The coatings in immersion service were checked every 6 months, while the panels under atmospheric exposure were inspected yearly. It was interesting to observe the widely differing service lives exhibited by the same generic types of coating under identical exposure conditions. In the case of the more

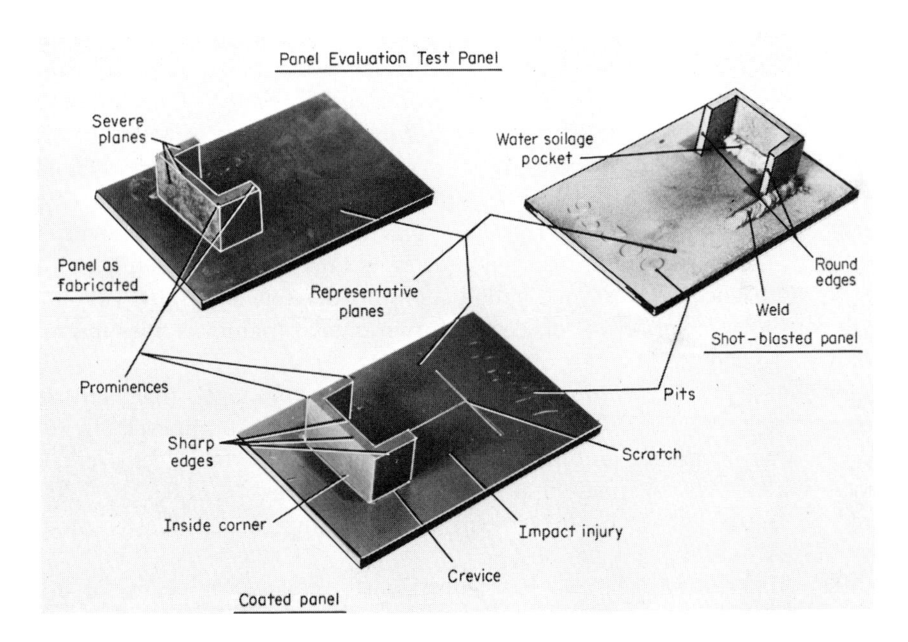

FIG. 6-1 Recommended paint panel for field exposure test. (*Kenneth Tator Associates, Coraopolis, Pennsylvania.*)

serviceable coatings, there was little difference between panels prepared by the manufacturer and those coated in the field. With less serviceable coatings, the field-prepared panels failed first.

Checking the panels for failure usually involves some form of visual inspection to determine the extent of undercutting corrosion, edge failure, rust spots per square inch, etc. The American Society of Testing Materials, Philadelphia, Pennsylvania, has been very active in establishing methods of paint testing of all kinds. They have available standards on conducting a panel exposure test program, and standards for rating paint failure due to rusting, blistering, chalking, cracking, flaking, erosion, etc. These ASTM Standards should be reviewed before planning any paint panel or patch-exposure program.

After selections are made based on composition and related serviceability or upon results of test exposure programs, the final choice is made based on considerations of ease of application, dry time, and material cost. The ease of application and waiting period between coats obviously affect application costs, and should be considered in making a choice between equally serviceable coatings. The cost per gallon is not so meaningful, however, and requires interpretation.

Material cost per gallon: The coverage obtained from a gallon of paint or coating depends upon its nonvolatile solids content. Theoretically, one gal of nonvolatile (100 percent solids) material will cover 1,604 sq ft at a thickness of 1 mil (0.001 in.). This generally is rounded off to 1,600 mil sq ft per gal for coverage on a smooth surface with no loss. If the paint contains 50 percent volatile solids, obviously only 800 sq ft would be covered with a 1-mil-thick dry film of paint.

The relationship between the wet paint cost in dollars per gallon, paint coverage and percent solids by volume, and the paint film cost per square foot for a 1-mil-thick dry film is presented in a monograph in Fig. 6-2. The paint material cost is obtained by connecting with a straight line the cost per gallon and the percent solids by volume. Obviously, a comparison of costs of two similar paints is meaningful only when compared on a material cost basis.

It should be cautioned that the theoretical average of 1,600 mil sq ft per gal is never fully obtained. Surface roughness appreciably reduces this coverage, and paint application losses further decrease the area which may be painted per gallon. A coverage of 1,000 to 1,200 mil sq ft per gal for a 100 percent solids coating on sandblasted steel would be considered good.

The above discussion of comparative costs is based on paints of the same generic type and composition. In many cases, differences in composition within the same type of paint also will have a marked effect on cost and quality. For example, the silicone content of silicone-alkyd paints will affect the heat resistance and cost of this type of paint. As another example, the serviceability (water resistance) and cost of chlorinated rubber paint are determined by the chlorinated rubber content of the vehicle.

When comparing different generic types of paints, cost per gallon and coverage have little meaning unless it is known that they will exhibit equal serviceability under the same exposure conditions. The representative current (1969) retail prices of different types of paints and coatings are presented in Table 6-2, and of solvents in Table 6-3. No professional or quantity purchase discounts have been included in these values. Paint costs would be reduced approximately 20 percent—and protective coatings 10 percent—over the 5-gal-can costs when purchased in 55-gal drums.

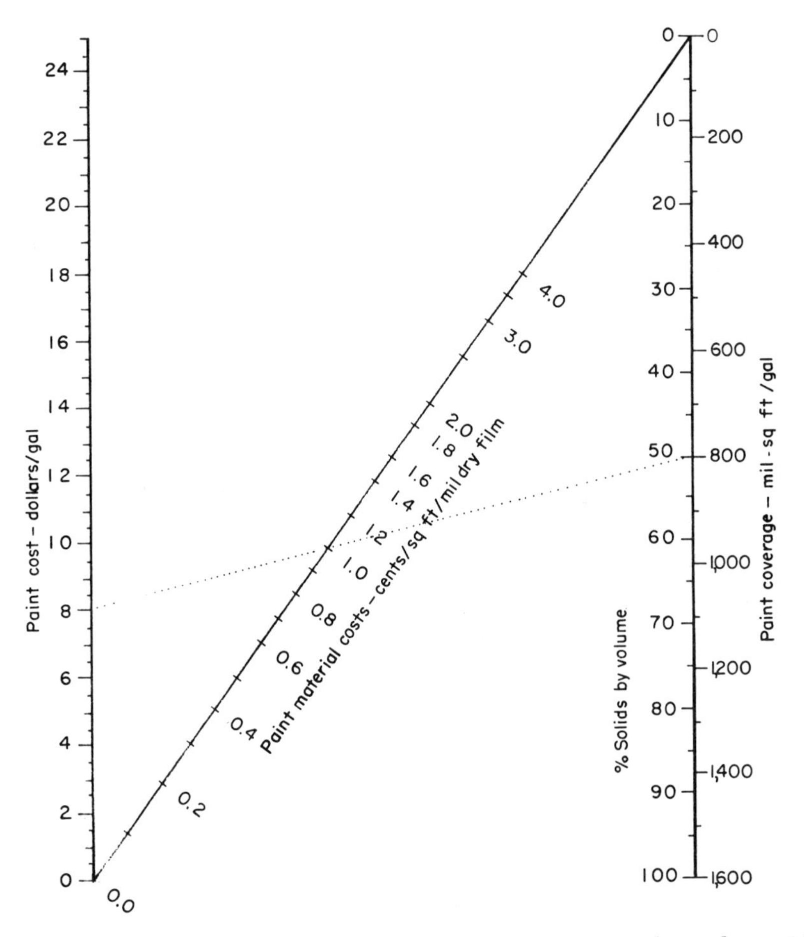

FIG. 6-2 Nomograph relating paint cost with coverage. (*Dow Chemical Co., Midland, Michigan.*)

TABLE 6-2 Representative 1969 Retail Prices of Paints and Coatings

Paints and coatings (by generic type)	*Approx. retail price, $/gal* (purchased in 5-gal pails)*
Drying oil (house paint)	8.00– 9.00
Alkyd	8.00–10.00
Phenolic varnish	9.00–12.00
Epoxy ester	6.00– 7.50
Chlorinated rubber (immersion)	6.00– 8.00
Vinyl—thin film (immersion)	5.00– 8.00
Vinyl acrylic	6.00– 8.00
Silicone	16.00–20.00
Silicone alkyd	9.00–15.00
Acrylic lacquer	10.00–13.00
Nitrocellulose lacquer	4.00– 6.00
Bituminous (cutback)	2.50– 3.00
Catalyzed epoxy	6.00–12.00
Catalyzed epoxy-phenolic	7.00–17.00
Catalyzed phenolic	15.00–25.00
Polyurethane (2-package)	14.00–18.00
Polyurethane (moisture cure)	5.00– 9.50
Coal-tar epoxy	6.50– 9.00
Hypalon	7.00– 8.00
Polyester (unpigmented, for FRP)	4.00– 8.00
PVA (water-base)	5.00– 8.00
Acrylic (water-base)	8.00–10.00
Butadiene-styrene (water-base)	6.50– 8.50
Zinc silicate primer (inorganic)	12.00–15.00
Zinc-rich primer (organic, catalytically cured)	12.50–16.00
Red lead alkyd primer	7.00–10.00
Zinc chromate alkyd primer	7.00–10.00

* No discounts are included in these values. Industrial maintenance paints and coatings almost always are sold at a discount, which ranges from 20 percent to as much as 50 percent off list price. House paints usually can be purchased at a 20 percent discount.

Surface Preparation Costs

The following discussion is concerned primarily with the cost of preparing steel surfaces for the application of maintenance paints and protective coatings. Surface-preparation methods considered are hand- and power-tool cleaning, and sandblasting.

Scraping and sanding of failed paint on wooden surfaces will be accomplished at widely varying rates depending upon the original condition of the surface. Consequently, averaged cost estimates such as the following should be used as guides only in estimating costs. The Painting and Decorating Contractors of America present in their "Estimating Guide" (1st ed., 1968) a rate of 185 sq ft per hr for sanding and puttying of plain siding and trim (100 sq ft per hr for trim only), and a rate

TABLE 6-3 Representative 1969 Retail Prices of Solvents

Type of solvent	Price, $/gal	
	5-gal pail	55-gal drum
Acetone	1.80	0.82
Alcohol, butyl	2.25	1.28
Alcohol, ethyl (95%)	2.05	0.93
Alcohol, isopropyl (91%)	2.00	0.85
Alcohol, methyl	2.75	0.65
Cyclohexanone	2.90	1.90
Ethylene glycol monoethyl ether	2.95	1.90
Ethylene glycol monoethyl ether acetate	2.90	1.80
Ethylene glycol monomethyl ether	3.05	1.95
Ethylene glycol monomethyl ether acetate	4.40	2.85
Methyl ethyl ketone (MEK)	2.10	1.15
Methyl isobutyl ketone (MIBK)	2.20	1.30
Mineral spirits	1.00–1.10	0.30–0.35
Naphtha, VM and P	1.10–1.35	0.31–0.38
Toluene (toluol)	1.40–1.50	0.44
Turpentine, gum spirits	3.00–3.25	1.30–1.40
Xylene (xylol)	1.50–1.60	0.48

Note: Refer to Table 3-4 on page 70 for trade names of the glycol ether solvents.

of 35 sq ft per hour for paint removal by burning on plain surfaces (25 sq ft per hr for siding).

The cost of preparing steel surfaces for painting also varies greatly depending upon the initial surface condition, the method of preparation employed, and the degree of cleanliness required. Martinson and Sisler reported the average 1960 costs of different methods of surface preparation ("Industrial Maintenance Painting: The Engineered Approach," Reinhold Book Corp., New York City, N.Y., 1961). These are presented below, together with a 1968 adjusted cost estimate based upon their 1960 figures.

Surface preparation method	Cost, dollars/sq ft	
	1960	Adjusted 1968 estimate
White metal blast	0.25	0.32
Commercial blast	0.13	0.17
Power tool	0.10	0.13
Hand clean	0.13	0.17
Steam/detergent clean	0.15	0.19

Surface-preparation cost estimates presented in the current marketing literature of a number of manufacturers of maintenance paints and coatings have been averaged with the following results:

Surface preparation method	*Cost, dollars/sq ft*
White metal blast	0.25–0.35
Commercial blast	0.11–0.16
Brush blast	0.05–0.08
Power tool	0.03–0.15
Hand clean	0.03–0.08

The surface-preparation cost estimates reported above are of interest for comparative purposes only. The hand- and power-tool cleaning costs are for the surface cleanliness produced by these methods, which of course is of a lower quality than produced by sandblasting. To produce a surface finish comparable to a commercial blast finish, hand cleaning—and to a lesser extent, power-tool cleaning—would require such extensive labor that these costs would be several times the cost of sandblasting.

TABLE 6-4 Average Rental Rates for Portable Air Compressors

Power source	Free air, cfm (at 100 psig discharge)	Average rental charges		
		Per month	Per week	Per day
Gasoline powered	20	$ 89.25	$ 31.25	$ 9.50
	36	90.75	32.75	9.95
	60	130.00	44.50	13.25
	75	149.00	50.75	15.50
	85	156.00	53.00	16.25
	100	169.00	56.75	17.25
	105	190.00	65.75	20.00
	125	210.00	71.00	22.50
Diesel powered	125	226.00	77.50	24.00
	150	242.00	80.50	26.25
	210	416.00	140.00	40.50
	250	429.00	146.00	44.25
	315	514.00	173.00	53.75
	365	572.00	194.00	63.25
	450	661.00	224.00	74.75
	600	858.00	293.00	98.75
	750	893.00	318.00	101.00
	900	1,234.00	426.00	140.00
	1,200	1,696.00	576.00	186.00

Extracted from the "1969 Compilation of Nationally Averaged Rental Rates," Associated Equipment Distributors, Oak Brook, Ill.

TABLE 6-5 Average Rental Rates for Sandblasting Equipment

Sandblast equipment	Average rental charges		
	Per month	Per week	Per day
Sandblast machines:			
Up to 100 lb inclusive	$ 86.25	$30.25	$10.50
101 to 250 lb inclusive	92.25	33.00	11.75
251 to 300 lb inclusive	122.00	45.50	14.50
301 to 600 lb inclusive	126.00	46.00	15.25
601 to 1000 lb inclusive	139.00	49.75	17.25
Hoses—complete with couples:			
¼ in. × 50 ft air hose	$ 7.50	$ 3.25	$ 1.30
¾ in. × 50 ft air hose	12.25	4.85	1.90
1 in. × 50 ft air hose	23.75	9.40	3.30
1¼ in. × 50 ft air hose	39.25	12.50	4.60
¾ in. × 50 ft sandblast hose	22.75	8.15	2.90
1 in. × 50 ft sandblast hose	30.75	11.50	4.40
1¼ in. × 50 ft sandblast hose	37.75	14.00	5.80
¾ in. × 10 ft whip hose	11.75	5.05	1.90
Hoods:			
Air fed	$ 27.00	$ 9.60	$ 3.95
Canvas	10.50	4.35	2.35
Nozzles:			
Short	$ 24.50	$ 8.95	$ 2.80
Long	26.75	9.20	2.95
Venturi style (short)	35.25	11.75	3.65
Venturi style (long)	39.00	13.25	4.75
Remote control "deadman" valve	50.50	19.25	6.60
Water trap	19.75	6.90	2.70
Wet blast head	13.70	4.70	1.80

Extracted from the "1969 Compilation of Nationally Averaged Rental Rates," Associated Equipment Distributors, Oak Brook, Ill.

Sandblasting is the principal method of surface preparation for the industrial application of maintenance coatings to steel surfaces. It is essentially the only suitable field method of preparation for coatings in immersion service. Consequently, sandblasting costs will be reviewed in somewhat greater detail.

Except for large plants where a continuous painting program is in effect, sandblast equipment usually is rented when needed. Current rental rates for air compressors are presented in Table 6-4, and for sandblast equipment in Table 6-5, as extracted from the "1969 Compilation of Nationally Averaged Rental Rates" by permission of the Associated Equipment Distributors, Oak Brook, Illinois.

Representative work rates for sandblasting steel surfaces under various

TABLE 6-6 Estimate of Sandblasting Rates and Sand Consumption
(Rates based on a ⅜ in. sandblast nozzle* using 1,400 lb sand/hr)

	White metal†		Commercial†		Brush-off†‡	
Surface condition	Rate, sq ft/hr	Sand, lb/sq ft	Rate, sq ft/hr	Sand, lb/sq ft	Rate, sq ft/hr	Sand, lb/sq ft
Loose mill scale; powdery rust.	125	11.2	250	5.6	500	2.8
Tight mill scale; little rust	90	15.5	200	7.0		
Coating blistered; hard rust scale; pits.	70	20.0	140	10.0	400	3.5
Badly pitted and rusted, with rust nodules.	. . .	. . . .	100	14.0		

* A ¼-in. nozzle and a 5⁄16-in. nozzle would clean at approximately 45 and 75% of the above rates, respectively. Sand consumption per sq ft would remain approximately the same.

† Refer to Chap. 4, page 92, for a description of surface preparation standards as defined by the Steel Structures Painting Council. "Near-white" blast cleaning rates are similar to "white metal" rates, varying from 0 to 20% higher for cleaning rate, and 0 to 20% lower in sand consumption per sq ft.

‡ "Brush-off" blast cleaning does not involve removal of tight paint or mill scale.

SOURCE: Painting and Decorating Contractors of America, Chicago, Ill.

initial conditions to a white metal, commercial, and brush-off surface finish are presented in Table 6-6, as estimated by the Painting and Decorating Contractors of America in their "Estimating Guide," 1st ed., 1968.

The total cost for a sandblasting operation consists of these factors:

Cost of equipment: Rental charges; or purchase costs, maintenance, and depreciation if owned

Cost of sand: Varies with locality; between $10 and $15 per ton delivered for clean river sand

Cost of labor: Also varies somewhat with locality; approximately $4 to $6 per hour

Cleaning rate: The big unknown; rates presented in Table 6-6 will serve as a guide. Many contractors estimate the cleaning rates for a job by first making a small scale sandblast test.

Combining the best estimates of cost and cleaning rates for sandblasting an "average" rusted surface to a near-white surface finish, it is determined in Table 6-7 that current sandblast costs are:

28 cents per sq ft for small jobs (¼-in. nozzle)
19 cents per sq ft for average industrial jobs (⅜-in. nozzle)
15.5 cents per sq ft for large industrial jobs (7⁄16-in. nozzle)

These costs were based on rental of all sandblast equipment on a daily rental rate basis. With equipment rental based on weekly rates, these costs become:

25 cents per sq ft for small jobs
17 cents per sq ft for average industrial jobs
14 cents per sq ft for large industrial jobs

TABLE 6-7 Estimate of Sandblasting Costs
(To provide a near-white surface finish)

For nozzle	Equipment sized for: small jobs (¼-in. nozzle) average jobs (⅜-in. nozzle) large jobs (7/16-in. nozzle)	Equipment rental/day (daily rates) Venturi nozzle size, in. ¼ in.	⅜ in.	7/16 in.
	Portable compressor*:			
¼ in.	125 cfm (gasoline)	$ 22.50		
⅜ in.	210 cfm (diesel)	....	$ 40.50	
7/16 in.	315 cfm (diesel)	....	....	$ 53.75
	Sandblast machine*:			
¼ in.	250 lb sand tank	11.75		
⅜ in.	600 lb sand tank	....	15.25	
7/16 in.	1000 lb sand tank	....	....	17.25
	50 ft hose, nozzles, hoods, etc.*:			
¼ in.	¾ in. air and sandblast hose; nozzle; hood	17.15		
⅜ in.	1 in. air and sandblast hose; nozzle; hood	....	20.85	
7/16 in.	1¼ in. air and sandblast hose; nozzle; hood	....	....	24.00
	Sand abrasive (6 hr/day blasting):			
¼ in.	600 lb/hr = 1.8 tons ($12/ton)	21.60		
⅜ in.	1400 lb/hr = 4.2 tons ($12/ton)	....	50.40	
7/16 in.	2000 lb/hr = 6.0 tons ($12/ton)	....	....	72.00
	Labor cost (8 hr/day labor):			
¼ in.	2 men; $48/day each	96.00		
⅜ in.	2½ men; $48/day each	....	120.00	
7/16 in.	3 men; $48/day each	....	....	144.00
	Total daily direct costs.........	$169.00	$ 247.00	$ 311.00
	Representative cleaning rates, and cost:			
¼ in.	100 sq ft/hr, 6 hr blasting	600 sq ft		
⅜ in.	220 sq ft/hr, 6 hr blasting	....	1,300 sq ft	
7/16 in.	320 sq ft/hr, 6 hr blasting	....	....	2,000 sq ft
	Cost/sq ft, cents...............	28.2	19.0	15.5

* Equipment rental rates based on the "1969 Compilation of Nationally Averaged Rental Rates," Associated Equipment Distributors, Oak Brook, Ill.

When the equipment is owned by the operator, the cost per square foot would be reduced still further.

It should be cautioned that although sandblasting is more expensive initially, it is the cheapest method in the long run for preparation of surfaces for maintenance coatings. It has been demonstrated many times that paints and coatings will exhibit two to three times the life when the initial application is made to a sandblasted surface as compared with a hand-cleaned surface.

Paint Application Costs

Application costs are primarily labor costs, and so any change that will reduce the labor time for application will be directly reflected in reduced application costs.

Labor time expended in application of paints and coatings will be dependent primarily upon three factors:

Method of application
Number of coats required
Nature of the structure being painted

The first two factors will be discussed briefly in this section. The paint applicator usually has little to say regarding structure design for improved coating application and protection against corrosion.

Method of application: As has been discussed in Chap. 5, "Paint Application," the most important and widely employed method of application of maintenance paints and protective coatings is by one of the spraying methods. However, application by brush or roller coater is the proper choice, and is widely employed, for application of the first primer coat to steel surfaces, for the application of paints and coatings to small areas, and in general for the application of house and interior wall paints.

At a painter's hourly wage rate of $8.00, which currently is in effect in many areas, the average application rates and costs for different methods of paint application to surfaces suited to that method are presented below:

TABLE 6-8 Application Rate and Cost Estimates

Application method	Application rate, sq ft/hr	Labor cost, cents/sq ft
Brush	150–250	3.2–5.3
Roller coat	200–300	2.7–4.0
Air spray	300–500	1.6–2.7
Airless spray	600–900	0.9–1.3
Air electrostatic	400–600	1.3–2.0
Airless electrostatic	800–1,100	0.7–1.0

These labor costs do not include time for paint preparation, scaffolding or positioning ladders, or for cleanup. These operations would increase the above costs by approximately 50 percent.

Application rates and paint coverage estimates based on actual experience are available for all house painting and most industrial painting operations from the Painting and Decorating Contractors of America. A very limited extraction of the extensive data available in their "Estimating Guide" (1st ed., 1968) is presented with their permission in Table 6-9.

A needless expenditure will be made for paint spray equipment if it is oversized for the application requirements, and a needless expenditure for excess painter's time will result if the equipment capacity is too small for the job. It is important that the equipment be matched to the job as far as possible.

For paint spray application to small areas and for touch-up painting, or where the scope of the paint job is such that the paint is purchased in 1-gal quantities or less: employ air spray guns equipped with siphon or pressure-feed cups.

For paint spray application to moderately sized areas of buildings or structures, especially when painting operations are not continuous, or where the scope of the paint jobs is such that paint is purchased in 5-gal pails: employ (1) air spray guns fed by pressure tanks when the air supply is limited, or by low-pressure pumps mounted on a 5-gal pail cover; or (2) airless spray guns fed by high-pressure pumps mounted on a cart (siphon pumps) or on a 5-gal pail cover.

For paint application to large areas in industrial applications, where the paint usually is purchased in 55-gal drums: employ air spray or airless guns fed by suitable pumps mounted on the 55-gal drum. The air or airless spray of heated materials should be considered for applications of this scope, and electrostatic spray equipment for applications to fences and open grillwork and for interior painting of structures and equipment.

Number of coats required: The service life of maintenance paints depends upon the thickness applied, all other factors equal. It is almost directly proportional to the mil thickness applied in excess of the anchor pattern depth of sandblasted surfaces.

In addition to considerations of thickness, the number of coats applied is important. Except for 100 percent solids materials, the evaporation of solvents will leave pinholes or pores in a paint film as it dries, and so a minimum of two coats is required—with three coats generally preferred.

The usual solvent-dry paints should not be applied thicker than 2 to 3 mils per coat (thickness depending upon the type of material applied), otherwise the film may crack, blister, or become spongy upon drying. High-build formulations applied in thick films per coat require

TABLE 6-9 Estimate of Work Rates and Paint Coverage

Surface preparation and painting Residence interior—representative operations	Work rate,* sq ft/hr	Coverage,* sq ft/gal
Trim, hand sanding:		
For paint, 1st coat	300	
For paint, incl. puttying	200	
For enamel, 1st coat	215	
For enamel, incl. puttying	115	
Wallboard, hand preparation: taping, bedding, sanding, etc	140	
Wallboard, machine preparation: taping, bedding, sanding, etc	200	
Painting and varnishing:		
Trim and windows:		
Brush (paint)	150–125	450–500
Brush (enamel)	150–90	450–500
Brush (stain)	220	500
Brush (shellac)	200	600
Brush (varnish)	175	600
Floors:		
Brush (enamel)	300	450–500
Roller coat (enamel)	400	430–575
Spray (enamel)	500	400
Brush (stainwax)	300	500–600
Roller coat (stainwax)	425	475–570
Wallboard, sandy plaster:		
Brush (texture coat)	150	100
Roller coat (texture coat)	275	45
Brush (water base)		
Roller coat (water base)	300	360
Spray (water base)	400–450	200–250
Brush (flat oil base)	225–175	300–350
Roller coat (flat oil base)	400–375	250–300
Smooth plaster:		
Brush (flat oil base)	225–175	400
Roller coat (flat oil base)	350–400	300–250
Acoustical surfaces:		
Brush (water base)	140	200
Roller coat (water base)	280	200
Spray (water base)	390	250

a dense thin-film seal coat applied either under or on top of the porous high-build coat.

Maintenance paints for exposure to mild atmospheres are recommended for application in 3- to 4-mils minimum total dry film thickness (2 mils above an anchor pattern height of 1 to 2 mils). In corrosive

TABLE 6-9 Estimate of Work Rates and Paint Coverage (Continued)

Surface preparation and painting Residence exterior—representative operations	Work rate,* sq ft/hr	Coverage,* sq ft/gal
Plain siding, incl. trim:		
Hand sanding	185	
Burning off paint	25	
Trim only:		
Hand sanding	100	
Flat surfaces:		
Burning off paint	35	
Masonry and brick:		
Wet sand blasting	500	
Paint application:		
Plain siding, incl. trim:		
Brush (oil base)	105–125	450–600
Roller coat (oil base)	150–200	400–550
Spray (oil base)	350–400	300–400
Brush (water base)	200	300
Roller coat (water base)	300	280
Spray (water base)	500	325
Trim and windows:		
Brush (oil base)	80–95	450–600
Doors and solid shutters:		
Brush (oil base)	250–125	400–350
Roller coat (oil base)	350–300	400–350
Spray (oil base)	500	400–450
Porch floors and steps:		
Brush (enamel)	175–225	375–550
Roller coat (enamel)	225–280	360–525
Spray (enamel)	350–400	350–400
Masonry and stucco walls:		
Brush (water base)	135–175	200–300
Roller coat (water base)	280–335	190–285
Spray (water base)	400–500	220–330
Brush (oil base)	75–95	200–300
Roller coat (oil base)	190–250	190–285
Spray (oil base)	350	220–330
Concrete floors and steps:		
Brush (enamel)	265–210	300–400
Roller coat (enamel)	340–360	285–380
Spray (enamel)	400–450	350–450
Gutters and downspouts:		
Acid wash	175	
Brush painting	75–100	250–300

* *Note:* The first figure is the work rate or coverage for applying the first coat; the second figure for subsequent coats (per coat).

Extracted from the "Estimating Guide," 1st ed., 1968, Painting and Decorating Contractors of America, Chicago, Ill.

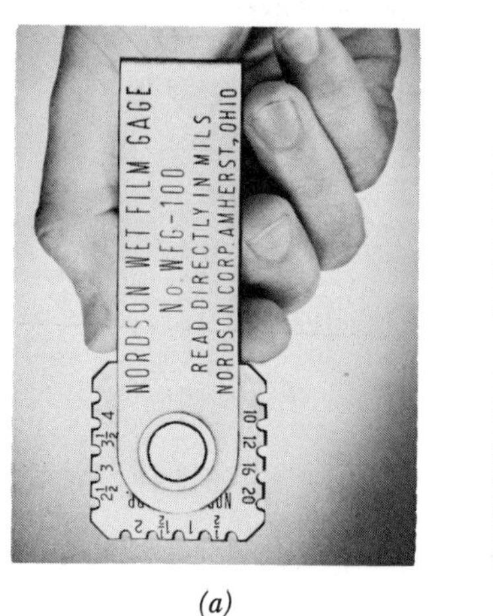

(a)

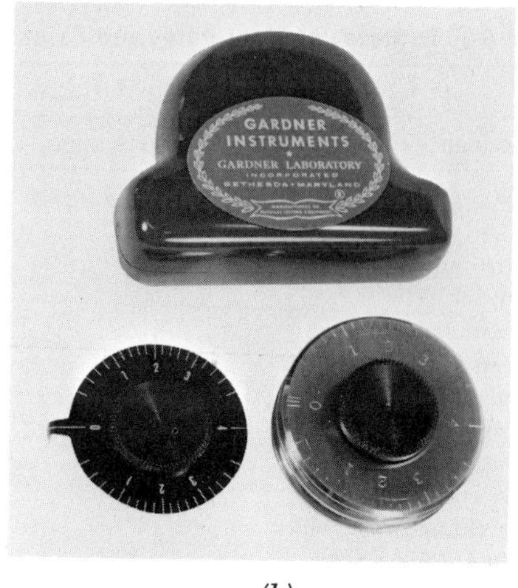

(b)

FIG. 6-3 Wet film thickness gauges. (*a*) **Nordson Wet Film Gauge** (*Nordson Corp., Amherst, Ohio*); (*b*) **Interchemical Gauge** (*Gardner Laboratory, Inc., Bethesda, Maryland.*)

or industrial atmospheres, 5- to 7-mils minimum total dry film thickness is recommended (4 mils above an anchor pattern height of 1 to 2 mils).

Heavy-duty coatings for immersion service or severely corrosive exposures are applied in thicknesses of 10 to 20 mils, depending upon the type of coating. Bituminous coatings are applied still thicker—30 to 60 mils ($\frac{1}{32}$ to $\frac{1}{16}$ in.) per coat.

In order to obtain the desired film thicknesses in as short a time and with as few coats as possible, hot spray application is recommended (either air spray or airless). Dry time between coats is reduced, and thicker films per coat may be applied with reduced porosity—all resulting in appreciable savings in labor costs and downtime losses.

Inspection: As coating application progresses, checks should be made on the thickness being applied. Simple but effective wet film thickness gauges are available for this purpose, as illustrated in Fig. 6-3.

The Nordson gauge is used by pressing the edge with the desired film thickness range squarely into the wet paint film and removing without sliding. The wet film thickness will lie between the thickness indicated on the highest step coated and the next higher step which did not contact the coating. The Gardner gauge is rolled over the wet coating. The film thickness is indicated on the rim of the roller opposite the point where the center roll first is contacted by the paint.

Small hand-held dry film thickness gauges are available, which determine film thicknesses by magnetic flux measurements. They are available in a number of film thickness ranges and can indicate film thicknesses with considerable accuracy. Some models are limited to measuring the thickness of nonmagnetic paint films applied to magnetic (ferrous) substrates as illustrated in Fig. 6-4a, b, and c, while others will work on nonmagnetic films when applied to nonferrous metals (aluminum, brass, copper, austenitic stainless, etc.), illustrated in Fig. 4d. More elaborate instruments, illustrated in Fig. 4e, will measure

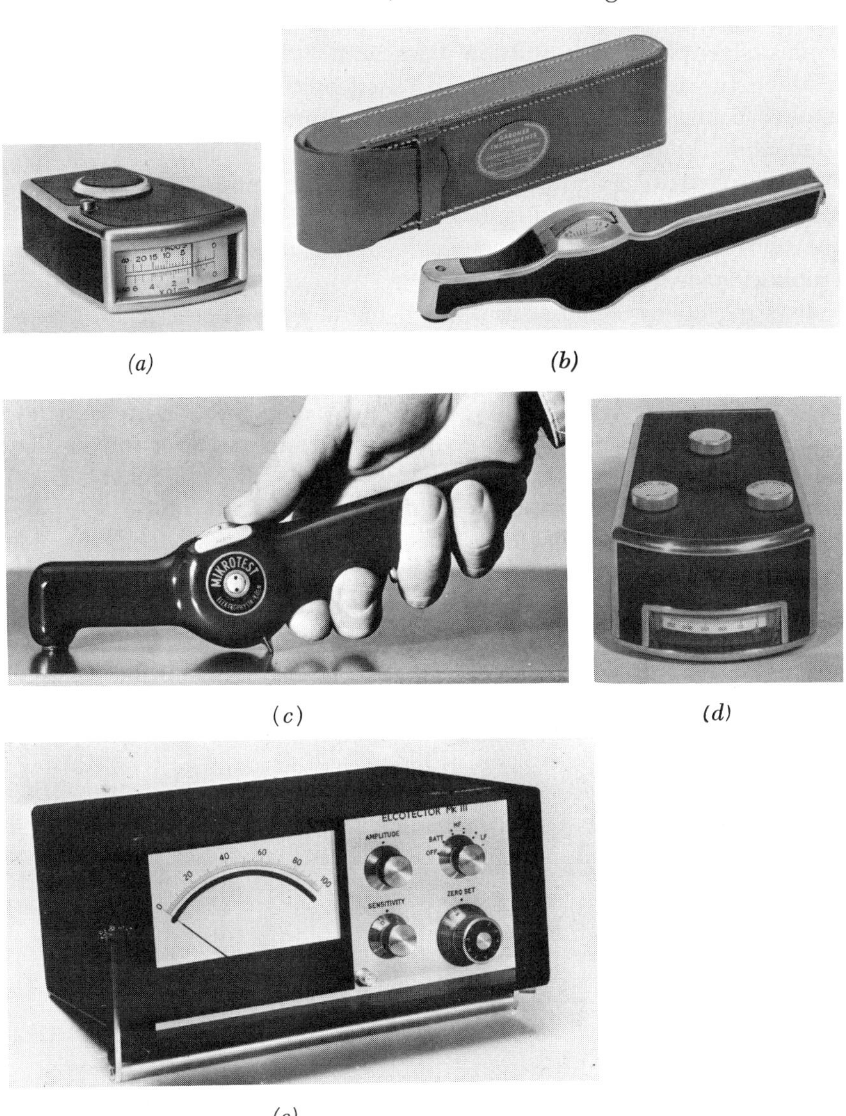

FIG. 6-4 Magnetic dry film thickness gauges. (*a*) **Elcometer Thickness Gauge.** (*Kenneth Tator Associates, Coraopolis, Pennsylvania.*) (*b*) **Inspector Thickness Gauge.** (*Gardner Laboratory, Inc., Bethesda, Maryland.*) (*c*) **Mikrotest Dry Film Gauge.** (*Kenneth Tator Associates, Coraopolis, Pennsylvania.*) (*d*) **Minitor Thickness Gauge.** (*Zormco Electronics Corp., Cleveland, Ohio.*) (*e*) **Elcotector Micro Gauge and Comparator.** (*Zormco Electronics Corp., Cleveland, Ohio.*)

the thickness of almost any type of film on any type of metallic substrate, as well as perform other measurements such as determining the hardness of steel, its alloy composition, location of minute cracks and flaws, etc.

Magnetic dry film gauges may give misleading film thicknesses of paint applied to rough or sandblasted steel surfaces—reading low by ½ to 1 mil. When calibration is necessary, they should be calibrated against steel plates at least ⅛ in. thick with similar surface roughness.

Magnetic dry film gauges can be used only for measuring the total applied paint film thickness when applied to metallic surfaces. In order to measure the thickness of the individual coats, or to measure paint films when applied to other than metal surfaces, an optical thickness gauge such as that illustrated in Fig. 6-5 can be used. A cut is scribed through the film, and measurements are made by means of a calibrated microscopic eyepiece.

It is essential that coatings applied for immersion service be free of all pinholes (holidays). The entire surface or, in the case of extensive coated areas, a large percentage of the total area should be checked by a holiday detector. In a typical holiday detector (Fig. 6-6), a low voltage is applied to a damp sponge which is moved over the surface. Wherever the coating is not continuous an electrical circuit is completed through to the metal substrate, and a signal operates.

Other models are available for checking pipe coatings. These employ coil spring electrodes which encircle the pipe. Adjustable high voltages are applied to detect any weak areas in the coating.

When poor adhesion between coats, or between the coating and the substrate is suspected or noticed, adhesion strengths should be checked. A rough check on adhesion may be made by cutting through the coating

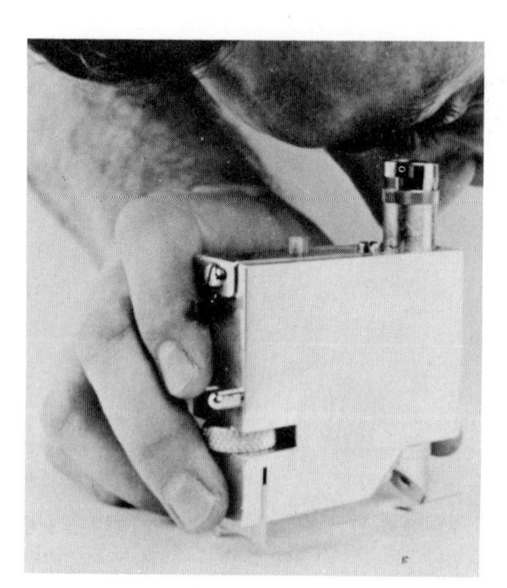

FIG. 6-5 Optical dry film thickness gauge. (*Kenneth Tator Associates, Coraopolis, Pennsylvania.*)

(a)

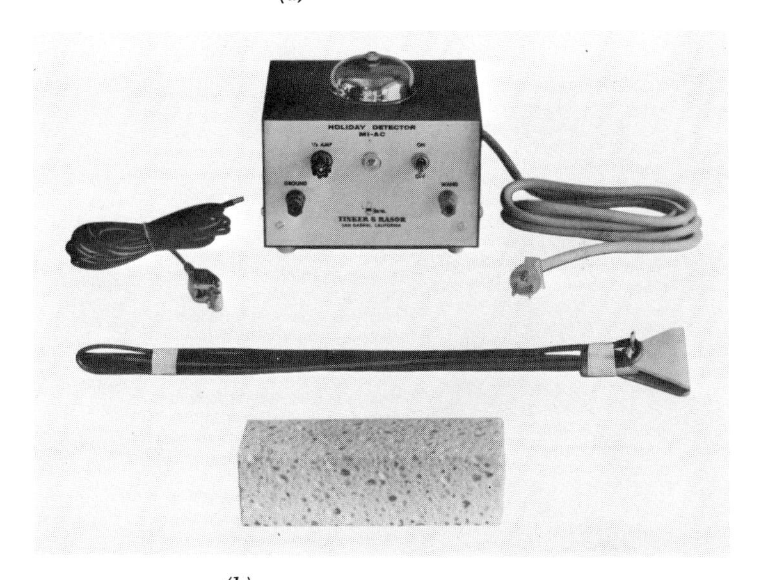

(b)

FIG. 6-6 Paint film holiday detectors. *(a and b—Tinker & Rasor, San Gabriel, California; c—Zormco Electronics Corp., Cleveland, Ohio.)*

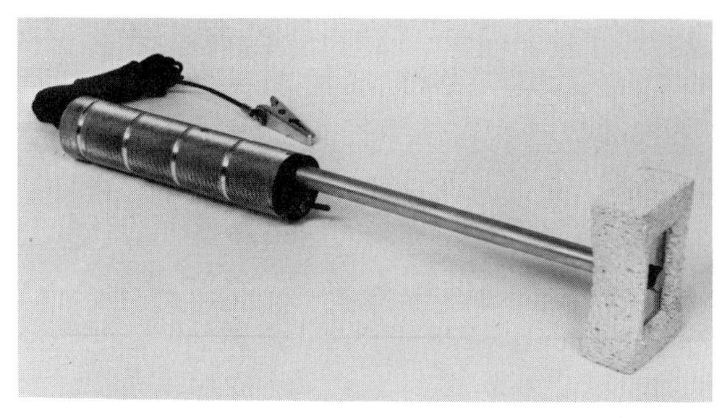

(*c*)

FIG. 6-6 (Continued)

with a knife and attempting to lift the film by inserting the knife blade under the film. A more exact and meaningful determination of adhesive strengths will be obtained through use of an adhesion tester such as illustrated in Fig. 6-7.

An aluminum button is cemented to the surface of the coating through use of a suitable adhesive. When the adhesive has cured, the claw of the instrument is attached and the wheel turned until the specified pounds of pull are reached—or until the button pulls off at some pull

FIG. 6-7 Adhesion tester. (*Zormco Electronics Corp., Cleveland, Ohio.*)

force below specifications. For paints meeting specifications, the button can be cut off and left on the paint film—painted over if desired.

The Applied Coating System and Its Maintenance

After having considered separately the individual costs of the principal factors involved in obtaining an applied coating system, the relationship between these separate costs now will be reviewed briefly. This will be followed by a few comments relative to a coating maintenance program and the paint labor forces employed.

Costs for the applied coating system: In the 1961 publication by Martinson and Sisler referred to earlier in this chapter, the 1960 breakdown of average costs for an applied protective coating was presented. These 1960 costs are presented below together with an estimate of 1968 costs based upon these figures:

Expense	Cost, cents/sq ft	
	1960	Adjusted 1968 estimate
Surface preparation	21	28
Material cost	8	10
Application	14	17
Miscellaneous costs	4	5
Total applied cost	47	60

In a survey of maintenance painting costs conducted in 1961 by the National Association of Corrosion Engineers (NACE Publication 61-14, Economics of Chemical Plant Maintenance Painting, *Corrosion*, Vol. 17, No. 12, December 1961) several interesting cost relationships were developed which still are of current interest. The relative costs of the operations required to apply a 3-coat maintenance paint system were found to be (average of 107 plants):

Expense	*Percentage of total applied cost*
Surface preparation	30 to 45% (average 40%)
Application (including material)	40 to 60% (average 48%)
Scaffolding, cleanup, miscellaneous	10 to 15% (average 12%)

Furthermore, the 1961 labor cost alone for surface preparation and application ranged between 70 and 77 percent of the total applied cost, with the remaining costs covering material and equipment.

A breakdown of total applied costs for a three-coat protective coating system developed from current cost information as presented in this chapter would be as follows:

Expense	*Cost, cents/sq ft*
Surface preparation (sandblast, near white)	30
Material cost ($10/gal, 300 sq ft/gal/coat)	10
Application (brush 1 coat; spray 2 coats)	10
Miscellaneous (scaffolding, paint mixing, cleanup)	5
Miscellaneous overhead costs	3
Total applied cost	58

Decorative and light-duty maintenance paints would be applied at a cost estimate of 40 to 60 percent of the cost for a protective coating: 25 to 35 cents per sq ft.

The cost figures presented above are of interest for comparison purposes only. Specific conditions vary too greatly for firm cost figures to be developed. Nevertheless, it is important to note that labor costs now comprise from 80 to 85 percent of the total applied cost. Consequently, any labor-saving improvement in material or its application will be directly reflected as savings.

Maintenance program: The successful and economical maintenance of a paint or coating system depends upon a regular inspection and touch-up program. The paint or coating seldom should be allowed to deteriorate to such an extent that it must be removed and reapplied.

Inspection of coatings in immersion service is dependent of course upon the frequency of downtime for test and inspection (T&I). These scheduled T&I periods usually are planned once every 12 to 24 or 30 months, depending upon service. During these downtimes the coating must be thoroughly checked and repaired. An unscheduled shutdown due to corrosion from a coating failure can be very costly.

A tank coating life, before recoating is necessary, may range anywhere between 5 and 10 years when properly maintained. Recoating should be scheduled for the next major T&I when inspection indicates the need.

Maintenance paints and coatings for atmospheric protection are not as dependent upon shutdowns for coating repair. A good program is to inspect and touch up a newly applied coating around 6 months after application to catch application faults, and then every 18 to 24 months thereafter.

A "heavy-duty" maintenance coating, properly kept in repair, should last from 8 to 12 years before recoating is necessary. Decorative and maintenance paints exhibit a shorter life—from 2 to 5 years. Often only one touch-up application is permitted because of the undesirable blotchy appearance that results from patch applications.

In general, when touch-up areas amount to 10 to 15 percent of the total area it is time to apply another coat or two. This will save the high costs of complete removal of failed paint and reapplication of the entire coating system. In the case of organic coatings applied over a zinc silicate primer, touchup, and reapplication of topcoat when necessary, should be so scheduled as to maintain protection of the zinc silicate. The zinc silicate will greatly extend the life of any organic topcoat as compared with the life of the same topcoat applied to organic primers on steel, as long as the expensively-applied zinc silicate coat is not allowed to deteriorate.

Touch-up costs per square foot of touch-up area are about double the initial application costs. However, since only 5 to 10 percent of the area should be involved, on a total area basis touchup will cost from 10 to 20 percent of first application costs.

An example of the estimation of the maintenance costs for a protective coating over a number of years is as follows:

ASSUME: Initial application cost: 50 cents per sq ft
Expected coating life: 10 years
4 touch-up applications required: 7 cents per sq ft

The total cost for 10-year protection will be 78 cents per sq ft or 8 cents per sq ft per year. Admittedly, this is a simplified example, presented on a cash basis. For proper cost estimating of a paint maintenance program, the present value of tomorrow's dollar must be considered.

Contract or plant force painting: The application and maintenance of industrial coatings usually are handled by a plant painting crew backed up by maintenance engineers, or the program may be contracted out on either a cost-plus or a firm price basis.

The National Association of Corrosion Engineers has made an interesting study of the advantages and disadvantages of each approach to maintenance painting (NACE Publication 6D168, Contract and Plant Force Painting: Advantages and Disadvantages, *Materials Protection,* Vol. 7, No. 2, February, 1968). The results of this study are presented in outline form below with their permission:

Plant Force Painting Advantages

1. Minimum supervision required
2. Minimum cost to prepare painting specifications
3. Minimum lost time due to changes in plans, schedules, specifications, etc.
4. Plant force painters usually develop a variety of painting skills, so they can efficiently handle a range of work from simple to complex
5. Pride in the job often results in high quality work

Plant Force Painting Disadvantages

1. Overhead is high since plant painters receive all fringe benefits of other employees

2. Payroll continues regardless of weather or workload

3. Efficiency and productivity reduced through higher absenteeism, longer work breaks, and by having older painters with long years of service in the paint crew

4. Inefficient or undesirable workers are difficult to replace

5. Plant painters require a plant investment in equipment

Cost-plus Painting Contract Advantages

1. Pride of workmanship leads to quality and efficiency

2. Due to job security considerations, efficiency is high

3. Flexibility exists in scheduling size or location of work force because of weather or workload

4. Labor costs are lower because of low overhead and fewer fringe benefits

5. Painters with specialized skills can be included in the crew

Cost-plus Painting Contract Disadvantages

1. Most of the risk is assumed by the plant owner

2. Jobs require considerable inspection and supervision

3. Detailed specifications must be prepared

4. Cost-plus painters are subjected to the same inefficiencies as plant painters: absenteeism, prolonged work breaks, etc.

5. Possible union conflicts may exist between the plant union and the contractor's union

Firm Price Contract Advantages

1. Most of the risk is assumed by the contractor

2. Jobs require no supervision, and inspection is necessary only to ensure specifications are met

3. Profit incentive leads to efficiency and high work output

4. No investment required in men or equipment, and other plant crafts seldom are involved

5. No involvement in purchase or storage of materials

Firm Price Contract Disadvantages

1. The contingency fee is higher because contractor assumes most of the risks

2. Detailed specifications are required, including such items as conduct of workers in the plant

3. Considerable detailed inspection is required, which should be agreed upon in advance

4. Changes in scheduling and workload due to plant operating requirements may result in claims for extra payments

Addendum: Specifications

In this addendum are presented a selected few of the 39 Paint System Specifications (PSS) and 20 Paint Application Specifications (PAS) which are available from the author.

Paint system specifications: The PSS present a number of detailed specifications for the application of selected examples of the more commonly employed paint and protective coating systems. They have been prepared for the convenience of the reader who plans to apply identical or similar types of paints and coatings. The complete listing of specifications available from the author is presented in Table A-1.

The Paint System Specifications are concerned primarily with details of application. They do not discuss the serviceability of the paint or coating under any particular exposure condition. (The selection of the type of coating recommended for a specific service condition is covered by the Paint Application Specifications, which follow.)

Each Paint System Specification is prepared in the same format and contains the following basic information:

1. Number of coats, dry time, and recommended dry film thickness
2. Safety precautions
3. Minimum surface preparation requirements
4. Instruction for preparation of the material for application
5. Application methods recommended
6. Characteristic properties of the applied coating and any special remarks

In the majority of cases, the specifications for protective coatings are written around specific proprietary coatings which have been found

TABLE A-1 Paint System Specifications (PSS)

PSS No. 1: Drying Oil Exterior Paint (Nonmetallic Surfaces)
PSS No. 2: Drying Oil Interior Flat Paint (Nonmetallic Surfaces)
PSS No. 3: Oil Modified Alkyd Semigloss Enamel (Nonmetallic Surfaces)
PSS No. 4: Oil Modified Alkyd Semigloss Enamel (Metallic Surfaces)
PSS No. 5: Short Oil Alkyd High Gloss Enamel (Nonmetallic Surfaces)
PSS No. 6: Medium Oil Alkyd High Gloss Enamel (Metallic Surfaces)
PSS No. 7: Medium Oil Alkyd High Gloss Enamel (Machinery and Equipment)
PSS No. 8: Oil Modified Alkyd Structural Steel Aluminum Paint
PSS No. 9: Oil Modified Alkyd Paint—Structural Steel Black
PSS No. 10: Exterior Phenolic Varnish
PSS No. 11: Epoxy Ester Enamel (Exterior Exposure)
PSS No. 12: Chlorinated Rubber Paint (Concrete and Wood)
PSS No. 13: Chlorinated Rubber Paint (Steel)
PSS No. 14: Nonskid Floor Paint
PSS No. 15: Vinyl-Acrylic Paint (Industrial and Humid Atmospheres)
PSS No. 16: Vinyl Mastic Coating (Industrial and Humid Atmospheres)
PSS No. 17: Vinyl Paint (Water Immersion)
PSS No. 18: Silicone Alkyd Paint
PSS No. 19: Silicone Aluminum High Temperature Paint
PSS No. 20: Nitrocellulose Lacquer
PSS No. 21: Acrylic Lacquer
PSS No. 22: Solvent Cutback Coal-tar Coating (Heavy Duty)
PSS No. 23: Amine-adduct-cured Epoxy Coating (Atmospheric Service)
PSS No. 24: Polyamid-cured Epoxy Coating (Immersion)
PSS No. 25: Catalytically Cured Phenolic Epoxy Coating (Immersion and Exterior Atmosphere)
PSS No. 26: Catalytically Cured Phenolic Epoxy Coating (Hot and Distilled Water Immersion)
PSS No. 27: Catalyzed Phenolic Coating (100% solids) (Heavy-duty Immersion)
PSS No. 28: Moisture-cure Polyurethane "Varnish" (Interior Wood)
PSS No. 29: Polyurethane Paint, 2 Package (Metal Surfaces)
PSS No. 30: Catalytically Cured Coal-tar Epoxy Coating
PSS No. 31: Hypalon (Chlorosulfonated Polyethylene) Coating (Heavy-duty Immersion)
PSS No. 32: Reinforced Polyester Coating (FRP)
PSS No. 33: Organic Zinc-rich Primer
PSS No. 34: Zinc Silicate Primer, Postcure Type (Steel Surfaces)
PSS No. 35: Self-cure Zinc Silicate Primer or Coating (Steel Surfaces)
PSS No. 36: Acrylic Water Emulsion Paint (Nonmetallic Surfaces)
PSS No. 37: Polyvinylacetate (PVA) Paint (Nonmetallic Surfaces)
PSS No. 38: Fire-retardant Paint
PSS No. 39: Fluorescent Alkyd Enamel

by actual panel exposure tests to provide excellent service under recommended exposure conditions. Like products from other manufacturers would be applied under closely similar specifications.

Grateful acknowledgment is accorded the following firms for their contribution of technical literature which proved helpful in preparing many of these specifications. The omission from this very brief list of the name of any of the approximately 1,700 paint and coating formulators now in the United States does not imply in any way that their product is not fully equal to those of the few listed firms.

Amercoat Corp., Brea, California
Carboline Co., St. Louis, Missouri
Cook Paint and Varnish Co., Kansas City, Missouri
Du Pont de Nemours & Co., Inc., Wilmington, Delaware
Fuller-O'Brien Corp., Menlo Park, California
The Glidden Co., Cleveland, Ohio
Koppers Company, Inc., Pittsburgh, Pennsylvania
Midland Industrial Finishes Co., Waukegan, Illinois
Napko Corp., Houston, Texas
PPG Industries, Pittsburgh Paint Div. Pittsburgh,, Pennsylvania
The Sherwin-Williams Co., Cleveland, Ohio
Wisconsin Protective Coating Corp., Green Bay, Wisconsin

Paint application specifications: The PAS present in concise form the recommended paint systems which may be applied to a specific structure, or for a specific service condition. They have been prepared for the convenience of the reader who is faced with the choice of a paint or coating for a certain condition. The complete listing of specifications is presented in Table A-2.

Since the number of different items painted is almost unlimited, only those items are included which are of major importance or which

TABLE A-2 Paint Application Specifications (PAS)

PAS No. 1: Interior Walls and Ceilings (Nonmetallic), including Wood Doors and Trim
PAS No. 2: Interior Clear Finished Wood and Plywood
PAS No. 3: Interior Sheet Metal Walls, Doors, and Frames
PAS No. 4: Exterior Wood—Siding and Trim
PAS No. 5: Exterior Natural Finished Woodwork
PAS No. 6: Concrete and Concrete Block Walls
PAS No. 7: Concrete Floors and Exterior Wood Porches and Platforms
PAS No. 8: Aluminum and Galvanized Steel
PAS No. 9: Structural Steel in Industrial Plants
PAS No. 10: Concrete Bases of Steel Supports and Structures
PAS No. 11: Metal at Elevated Temperatures: 500°F Maximum
PAS No. 12: High Temperature Surfaces: above 500°F (including Furnace Stacks)
PAS No. 13: Storage Tank—Exterior
PAS No. 14: Oil Storage Tank—Interior Bottom
PAS No. 15: Water Tank Interiors
PAS No. 16: Storage Tank Interiors for Chemicals, Solvents, etc.
PAS No. 17: Water Softener and Demineralization Plants
PAS No. 18: Swimming Pool Filters
PAS No. 19: Marine Structures: Atmospheric Exposure
PAS No. 20: Water-cooling Towers

present unusual conditions. Special applications for which specified brands of materials are used such as bowling alleys and pins, gymnasium floors, etc., are not included in these specifications.

The Paint Application Specification (PAS) for each structure or application condition covers the following subjects:

1. Scope: The scope and limitations of the application are specified, especially with respect to the surfaces to be painted.

2. Areas to be Painted: The PAS specifies areas to be painted, with the understanding that the project scope of work, work order, or other authority may supersede these instructions for reasons which lie outside the scope of this presentation.

3. Surface Preparation Requirements: In many cases, normal surface preparation methods are mentioned without much detail. Where the surface preparation is critical, more elaborate requirements or recommendations are given. However, in individual cases where the conditions may vary considerably, judgment and skill based on the principles of good workmanship should supplement the instructions presented in the "Applications Manual for Paint and Protective Coatings."

4. Paint System to be Applied: The PAS specifies the paint systems to be applied to the various areas. Multiple choices are mentioned where possible. The minimum total dry film thickness or number of coats is specified with the paint system either in this PAS or by reference to the Paint System Specification (PSS).

5. Color Code or Color Requirements: If a few colors only are involved, they are specified in the PAS. Otherwise, reference is made to Chap. 2, where color requirements are covered in greater detail. This is done to avoid excessive repetition.

6. Safety Precautions: Any safety precautions or any health safeguard which is more or less special for the item to be painted is given in the applicable PAS. Also, the more common safety instructions are listed as a reminder to the man on the job even though they are discussed more completely in Chap. 3.

In summary, these specifications present such subjects as the recommended choice of paint to be applied, areas to be painted, and colors, while the Paint System Specifications specify how this recommended paint is to be prepared and applied.

Three representative Paint System Specifications and Paint Application Specifications are presented on the following pages. The complete set of these specifications as listed in Tables A-1 and 2 may be obtained at a nominal charge by contacting the author: William F. Gross, Binks Research and Development Corporation, 1791 Range Street, Boulder, Colorado 80301.

PAINT SYSTEM SPECIFICATION NO. 3
Oil Modified Alkyd Enamel
Semigloss Enamel System for Nonmetallic Surfaces

Coating System to Be Applied

1. To exterior surfaces, apply:
One coat primer-sealer and two or more coats semigloss enamel as required.
2. To interior surfaces, apply:
One coat primer-sealer, one or two coats enamel undercoater, and one coat semigloss enamel.

The number of topcoats also depends on the desired color. Light tints and yellow and red enamel will not always cover any contrast in one single coat.

Safety Precautions

The solvents and thinners are flammable and moderately toxic. The operator engaged in surface preparation and paint application must use the protective equipment and apparel indicated in Chap. 3.

Surface Preparation Limitations

The primer-sealer is a controlled penetration-type sealer suitable for most porous surfaces, provided the surfaces are dry, clean, neutral, and free of loose or flaky matter.

TABLE A-3

Material	Application method	Coverage, sq ft/gal	Wet film thickness	Dry time to recoat	Dry film thickness
Primer-sealer, white	Brush	300	3 mils	24 hr	1 mil
Enamel undercoat, white	Brush or spray	350	3 mils	18 hr	1½ mils
Semigloss enamel	Brush or spray	400	2½ mils	24 hr	1 mil
Pigments-in-oil (various colors)	Use to tint enamel undercoater and enamel to the desired pastel shade				
Drier	Use to restore proper drying qualities				
Mineral spirits	Equipment cleaner and thinner (aliphatic)				
Enamel thinner	Aromatic thinner for enamels				
Turpentine	Thinner for paints and enamels				

Concrete and plaster must be fully cured and aged. Thirty days is considered a minimum dry time for plasterwork under good drying conditions. For concrete, at least three months are required before painting. Fresh concrete and plaster may be pretreated with dilute acids, zinc sulphate, or zinc chloride solutions.

A water-base PVA seal coat may be applied to concrete or plaster not fully cured, after which the oil-base paint system may be applied without waiting for full cure of the concrete.

Exterior woodwork preferably shall have a moisture content between 9 and 14 percent. Interior woodwork preferably shall have from 5 to 10 percent moisture.

Normal surface preparation methods for wood, board, asbestos cement, concrete, plaster, and other nonmetallic surfaces are described in Chap. 4.

Preparation of Paints

Agitate or stir paint and enamel until uniform. Do not add thinner for brushing. For spraying add up to one pint of thinner to one gal of paint or enamel as required to obtain proper consistency. Use thinner sparingly.

If paint of a particular batch dries too slowly, add up to, but not more than, one pint of drier to each five gallons of paint. Mix well. Excessive amounts of drier may ruin the paint.

Mix the enamel undercoater with pigments-in-oil to approximate the color of the finish coat. Use up to about 5 percent by volume, but never more than 10 percent of pigments-in-oil to color the semigloss white enamel to the desired light tint. Mix the pigments uniformly with the white base. Deep or strong colors cannot be prepared in this manner and must be pigmented to the desired color as received.

Application Conditions

Apply the first coat as soon as possible after surface preparation. New exterior woodwork should be finished as soon as practical after erection.

Apply primer-sealer by brush. Brush well into pores and voids for maximum bond. Allow primer to dry hard.

For exterior and interior woodwork, where a good appearance is required, fill nail holes, cracks, and joints with putty after priming and sand smooth. Remember that the semigloss enamel may accentuate defects or irregularities in the surfaces.

Undercoater and enamels may be applied by brush or spray. Allow each coat to dry completely before applying the next coat.

Conventional spray application conditions, for both primer-sealer and top coats:

FLUID PRESSURE: 25 to 30 psig
ATOMIZING AIR PRESSURE: 30 to 40 psig
FLUID ORIFICE SIZE: 0.070 to 0.086 in.
AIR NOZZLE AIR RATE: 12 to 14 cfm at 50 psig

Airless spray application conditions:

FLUID PRESSURE: 1,600 to 2,000 psig
FLUID ORIFICE SIZE: 0.015 to 0.018 in.

Sand undercoater with waterproof sandpaper and water when a very smooth finish is required. In this case two coats of undercoater are recommended.

Notes

This paint system is intended for exterior and interior nonmetallic surfaces, where a semigloss finish is desired. It has good weathering properties and withstands washing well. However, there may be several complex causes for locally premature failure or deterioration of the paint system. One major cause is related to entry of moisture behind the paint coating. Another common cause is the excessive coating thickness built up by frequent repainting.

Chapter 5 contains more detailed information on types and causes of paint failure.

PAINT SYSTEM SPECIFICATION NO. 12

Chlorinated Rubber Paint System for Concrete and Wood Surfaces

Coating System to be Applied

1. To unpainted porous surfaces, apply:

One coat of thinned chlorinated rubber paint, plus two or more full coats of chlorinated rubber paint.

2. To painted surfaces, apply:

Two or more coats of chlorinated rubber paint.

Determine the number of coats from the table below to obtain the total dry film thickness specified in the applicable Paint Application Specification.

TABLE A-4

Material	Application method	Coverage, sq ft/gal	Wet film thickness	Dry time to recoat	Dry film thickness
Chlorinated rubber paint (various colors)	Brush	300	3½ mils	24 hr	1½ mils
	Spray when thinned	400	2½–3 mils	24 hr	1 mil
Enamel thinner	High aromatic solvent (slow evaporation)				
Xylene	Thinner and equipment cleaner				
Toluene	Thinner and equipment cleaner (rapid evaporation)				
MIBK	Use in spraying thinner, below				
Spraying thinner	Add 1 part MIBK to 3 or 4 parts xylene or toluene				

Safety Precautions

The solvents in chlorinated rubber paints as well as the thinners and equipment cleaner are toxic and flammable. When coating is applied in confined areas, adequate ventilation is required to keep the fumes below the lower flammable and toxic limits.

Avoid prolonged breathing of vapors or prolonged contact with the skin. The operator engaged in surface preparation and paint application must use the protective equipment and apparel indicated in Chap. 3 of this manual.

Surface Preparation Limitations

The chlorinated rubber paints have poor wetting properties, and therefore require a dry, well-prepared surface.

Concrete surfaces must be completely dry and free of glaze, salt deposits, and loose or unsound matter. Woodwork shall preferably have a moisture content not over 14 percent. If the moisture content is higher, adhesion will be impaired.

Normal surface preparation methods for wood and concrete are described in Chap. 4.

Preparation of Paints

Agitate or stir thoroughly until uniform. Repeat stirring at regular intervals during application.

Thin material for the first coat applied to bare porous surfaces with one part of enamel thinner to four parts of paint.

For spray application add up to 15 percent by volume of spraying thinner, as required to obtain optimum spraying properties. Chlorinated rubber paints are very difficult to spray because of cobwebbing.

Never use more thinner than is absolutely necessary. The spraying thinner specified above has a high solvency power and will effectively reduce the viscosity, even when used in small amounts.

Application Conditions

Apply the first coat with a full brush, working from uncoated to coated areas. Brush well into the surface, but do not rebrush the partly dried areas.

Subsequent coats of chlorinated rubber paint may be brushed or sprayed. Allow each coat to dry completely.

When brushing subsequent coats of chlorinated rubber paint, use a full brush all the time and "flow" the material onto the surface. Excessive brushing will cause lifting of the previous coats.

Spraying of chlorinated rubber paints requires a careful adjustment of paint viscosity and of atomizing pressure in order to avoid cobwebbing, and to obtain a continuous wet film free from pinholes. Only skilled spray applicators can successfully apply chlorinated rubber paints by spraying. In general, spray application of chlorinated rubber paints is not recommended. The following conditions apply to coatings especially formulated for spraying.

Conventional spray application conditions

FLUID PRESSURE: 8 to 12 psig

ATOMIZING AIR PRESSURE: 50 to 65 psig

FLUID ORIFICE SIZE: 0.070 to 0.078 in.

AIR NOZZLE AIR RATE: 12 to 14 cfm at 50 psig

Airless spray application conditions

FLUID PRESSURE: 2,000 to 2,500 psig

FLUID ORIFICE SIZE: 0.011 to 0.015 in.

If the surface will be exposed to chemical corrosion but not to traffic or abrasion, select different colors for consecutive coats to assist in gauging the coating thickness.

Notes

Chlorinated rubber paints are resistant to mineral oils and to solutions of most acids, alkalies, and salts. They are also highly resistant to humidity, condensation, and wetting with salt or fresh water. They are not resistant to vegetable oils and to aromatic or chlorinated solvents.

The dried film is impermeable to moisture. Consequently, the surface must be dry during application. Any moisture or dampness will be trapped under the coating and will cause blistering.

Chlorinated rubber paints are flame retardant and abrasion resistant. The maximum temperature for continuous exposure is approximately 140°F.

This paint system is recommended for concrete bases and supports, concrete floors, wooden decks and porches, and for swimming pools.

PAINT SYSTEM SPECIFICATION NO. 23
Amine-adduct-cured Epoxy Coating System
Atmospheric Exposure

Coating System to be Applied

To a sandblasted or power tool-cleaned surface, apply:

One coat of inhibitive epoxy primer, and two coats catalyzed epoxy top coat.

Determine the number of coats from the table below to obtain the total dry film thickness specified in the applicable Paint Application Specifications.

TABLE A-5

Material	Application method	Coverage, sq ft/gal	Wet film thickness	Cure time to recoat	Dry film thickness
Epoxy red lead primer	Brush	300	2½ mils	4 hr	1½ mils
	Spray	300	3 mils	4 hr	2 mils
Catalyzed epoxy top coat (various colors)	Spray	150	8 mils	8 hr	4–5 mils
MIBK (50%) Toluene (50%)	Equipment cleaner and thinner				

Safety Precautions

The solvents in catalyzed epoxy coatings as well as the thinners and equipment cleaner are toxic and flammable.

The primer employs a polyamid catalyst which, although only mildly toxic, should be washed from the skin when contacted, to prevent a skin rash. The amine-adduct catalyst used with the topcoat is relatively nontoxic.

When coating is applied in confined areas, adequate ventilation is required to keep the fumes below the lower flammable and toxic limits. Avoid prolonged breathing of vapors or prolonged contact with the skin.

The operator engaged in surface preparation and paint application must use the protective equipment and apparel indicated in Chap. 3.

Surface Preparation Limitations

The epoxy primer shows good adhesion to nearly all types of surfaces provided they are dry and free of grease and oil. Depending upon the corrosive degree of the service atmosphere, surface preparation may be by good quality power-tool cleaning, or preferably by sandblasting to a "brush-off" or "commercial gray" finish.

The primer should be applied soon after surface preparation to minimize contamination before priming.

Preparation of Paints

1. Primer: The primer is a polyamid-catalyzed epoxy primer containing inhibitive (red lead/iron oxide) pigments. One part by volume of the catalyst is added to four parts epoxy resin base. After stirring both containers separately, the catalyst is slowly added to the resin base with constant stirring, continuing for several minutes. Pot life of the catalyzed primer is 8 to 10 hr at 70 to 80°F.

2. Topcoat: The topcoat is an amine-adduct-cured epoxy coating. One part by volume of amine-adduct catalyst is added to two parts epoxy resin base with thorough stirring. Pot life of the catalyzed coating is 4 hr at 70 to 80°F.

No thinning normally is necessary for application either with the primer or top coat.

Application Conditions

1. The primer may be applied by brush or spray methods. Spraying usually is preferred. Spray full wet coats with 50 percent overlap. Two parallel passes should be sufficient to apply the desired 3- mils wet film thickness.

Conventional spray application conditions

FLUID PRESSURE: 15 to 25 psig

ATOMIZING AIR PRESSURE: 50 to 70 psig

FLUID ORIFICE SIZE: 0.059 to 0.070 in.

AIR NOZZLE AIR RATE: 13 to 16 cfm at 50 psig

Airless spray application conditions

FLUID PRESSURE: 2,000 to 2,500 psig

FLUID ORIFICE SIZE: 0.015 to 0.018 in.

2. The topcoat is spray-applied except for small area touch-up by brush. Apply full wet coats by a cross hatch technique.

Conventional spray application conditions

FLUID PRESSURE: 25 to 40 psig

ATOMIZING AIR PRESSURE: 60 to 80 psig

FLUID ORIFICE SIZE: 0.059 to 0.070 in.

AIR NOZZLE AIR RATE: 13 to 16 cfm at 50 psig

Airless spray application conditions

FLUID PRESSURE: 2,000 to 2,500 psig

FLUID ORIFICE SIZE: 0.015 to 0.018 in.

Notes

The aminc-adduct-cured epoxy coating system provides superior protection to steel surfaces under highly corrosive atmospheric exposure conditions. Protection is afforded against splash and spillage of water, oils, salts, acids, alkalies, and other corrosive chemicals.

PAINT APPLICATION SPECIFICATION NO. 4

Exterior Wood—Siding and Trim

Scope

This specification covers the painting of exterior wood and plywood siding, doors (unless varnished), door frames, window frames, and other wood trim.

Areas to Be Painted

Paint all exposed exterior wood and plywood as mentioned above.

Surface Preparation Requirements

New wood shall be planed and/or sanded smooth and clean. For best results the wood should have a moisture content between 9 and 14 percent. Plywood shall be exterior type and any dirt or markings shall be removed.

Previously painted wood and plywood shall be scraped or wire brushed to remove any blistered, loose, or cracked paint. Edges of intact coating should be sanded to make a smooth surface. If the surface is otherwise reasonably clean and dry, brushing with a painter's duster may suffice for most of the area. Very dirty and greasy parts should be washed. Contamination resulting from asphalt spilled in roof repairs or resin exuded from the wood should be removed with scrapers and solvents.

Complete removal of the old paint, although a costly operation, is sometimes necessary. This will be the case when the coating thickness has become excessive from frequent repainting, causing dry blistering and conspicuous scaling and curling, or when the cracking has lost all relation to the grain pattern of the wood.

It is advisable to remove coatings more than 10 mils thick, and coatings 20 mils or more in thickness definitely must be removed. This should be done by paint remover, scraping, and washing. Burning off old paint is a dangerous process and should be attempted only by a painter experienced in this process.

Surfaces showing abnormal deterioration of the paint should not be repainted before the causes of the abnormal behavior have been carefully examined and, if possible, corrected. See Chaps. 1 and 4.

Paint Systems to Be Applied

1. On siding: Exterior Oil Paint System for Nonmetallic surfaces, PSS No. 1
2. On doors, door frames, roof trim: High Gloss Enamel System for Nonmetallic Surfaces, PSS No. 5 or
3. Semigloss Enamel System for Nonmetallic Surfaces, PSS No. 3.

Apply the minimum number of coats for full hiding—usually two or three.

Color Code or Color Requirements

No color requirements apply.

Safety Precautions

Observe the safety precautions given in the applicable Paint System Specifications. These precautions are quite ordinary safety measures and should be part of the painter's safety habits. They are important enough to be kept in mind at all times, however.

Notes

When paint is applied and maintained in accordance with a well-planned program, the surface preparation for repainting is normally relatively easy and inexpensive. The paint should be given time to waste away materially between paintings; otherwise the coating soon becomes too thick and must be removed. Therefore, the rate of normal deterioration of the paint should govern the frequency of repainting. On the other hand, a satisfactory appearance during prolonged periods between paintings requires good surface preparation and application methods.

PAINT APPLICATION SPECIFICATION NO. 7

Concrete Floors and Exterior Wood Porches and Platforms

Scope

This specification covers the painting of exterior or interior concrete floors, and exterior wood porches, platforms, and steps.

Areas to Be Painted

New and existing concrete floors shall be painted only on special order or when required by the project scope of work (see Notes). Concrete floors and curbs shall be marked to set sections aside for such reasons as traffic, storage areas, and dangerous locations in accordance with the Safety Color Code presented in Chap. 2.

All exposed surfaces of wood porches, platforms, steps, including handrails and other appurtenances which are part of permanent or semipermanent buildings, shall be painted.

Surface Preparations Requirements

Normal surface preparation methods for concrete and wood as described in Chap. 4 on surface preparation shall be followed.

The surface shall comply with the limiting requirements for surface condition as contained in the applicable Paint System Specifications.

Paint System to Be Applied

1. For concrete and wood floors and wood trim subjected to normal traffic and wetting:

a. Chlorinated Rubber Paint System for Concrete and Wood, PSS No. 12. Apply not less than three coats.

b. Epoxy Coating System, Amine-adduct-cured, PSS No. 23. (Omit the metal primer.) Two coats are recommended for wear resistance and life.

Where slippery conditions may be present, apply one additional coat of Nonskid Floor Paint System, PSS No. 14. For lines and marking, use either zone-marking lacquer or chlorinated rubber paint in the color required.

2. For floors in chemical plants and other very corrosive areas, select one of the following coating types:

a. Polyurethane Paint System (two-package), PSS No. 29. (Omit the metal primer.) Apply one base coat and one seal coat.

b. Catalytically cured Phenolic Epoxy Coating System, PSS No. 25. (Omit the metal primer.) Apply at least two coats for wear life.

c. Catalyzed Phenolic Coating System, PSS No. 27. (Omit the metal primer.) Apply two thin coats to a total dry film thickness of 8 to 10 mils.

d. Hypalon Coating System, PSS No. 31. Apply one coat each of seal, body, and clear finish coat.

Color Code or Color Requirements

Standard floor colors are light gray, medium green, and tile red. The color may be selected from these to blend with the general color scheme. Other colors also are available.

Lines and markings shall have the color specified by the Safety Color Code presented in Chap. 2.

Safety Precautions

Observe the safety precautions given in the applicable Paint System Specifications. These precautions are quite ordinary safety measures and should be part of the painter's safety habits. They are important enough to be kept in mind at all times.

Notes

Although concrete floors often are left unpainted, it may be desired that they be sealed to reduce dusting for easier cleaning, etc. Two or three applications of boiled linseed oil diluted $\frac{1}{3}$ to $\frac{1}{2}$ with mineral spirits make an effective concrete floor sealer for most purposes. Proprietary silicone-base concrete floor sealers also are recommended.

PAINT APPLICATION SPECIFICATION NO. 9

Structural Steel in Industrial Plants

Scope

This specification covers the painting of exterior, open, or sheltered steel structures in industrial plants and refineries.

Areas to Be Painted

All exterior exposed structural steel at ambient temperatures:

1. Supports and frames for piping and equipment
2. Plant buildings and shelters
3. Platforms, including floor plates, grating, stairs, handrails, toeboards
4. Light poles, masts, fences, gates, posts, markers, and other miscellaneous items

Surface Preparation Requirements

Selection of the most suitable method of surface preparation is dependent on the location, nature, accessibility, and surface condition of the object to be painted. The prepared surface must comply with the limiting requirements of the paint system to be applied. See applicable Paint System Specifications.

Whatever method is selected, it must be carried out in a good workmanlike manner. For detailed information on surface preparation methods, refer to Chap. 4.

Remove old paint coats which have become excessively thick after many years of repainting and have started to crack and scale. This may be accomplished by chipping and descaling, followed by sandblasting if permitted.

Paint System to Be Applied

1. For mild industrial and refinery exposure conditions, select paint system to be applied from those listed below, depending on exposure condition and finish color. (Also, see Notes below.)

a. Structural Steel Aluminum Paint System, PSS No. 8. For black bases of supports and frames, apply one coat of structural steel black paint over the aluminum paint system.

b. Structural Steel Black Paint System, PSS No. 9.

c. Vinyl-acrylic Paint System, PSS No. 15.

d. Epoxy Ester Paint System, PSS No. 11.

e. Semigloss Enamel System for Metal Surfaces, PSS No. 4.

f. High Gloss Enamel System for Metal Surfaces, PSS No. 6.

Minimum total dry film thickness for all above systems: 5 mils. When repainting, avoid building up unduly thick coatings.

2. For corrosive atmospheres, such as in a chemical plant, select one of the coating systems listed below:

a. Chlorinated Rubber Coating for Steel, PSS No. 13. Apply two coats of primer and two coats of topcoat, for a total dry film thickness of 5-mils minimum.

b. Catalytically Cured Phenolic Epoxy Coating, PSS No. 25. Apply one coat of primer and two coats of topcoat, for a total dry film thickness of 6 to 8 mils.

c. Vinyl Mastic Coating System, PSS No. 16. Apply two coats of primer and one coat of topcoat, for a total dry film thickness of 6 to 8 mils.

d. Epoxy Coating System, Amine-adduct-cured, PSS No. 23. Apply one coat of primer and one coat of topcoat, for a total dry film thickness of 6 to 8 mils.

e. Polyurethane Paint System (two-package), PSS No. 29. Apply one coat each of primer, base coat, and seal coat, for a total dry film thickness of 7 to 8 mils.

3. For very severe and corrosive exposure conditions, first apply a zinc-rich primer (PSS No. 33) or zinc silicate primer (self-cure type preferred—PSS No. 35). The primer then may be seal-coated by any of the coating systems listed in (2), above (omitting the primer).

Color Code or Color Requirements

Observe the safety color code presented in Chap. 2, and mark locations which may create a physical hazard. For example, paint yellow and black stripes on the parts of a structure that may form an obstruction to traffic of any kind; paint general plant structural steel white, gray, or aluminum; paint a black 6-ft-high base on pipe supports and frames; paint stairs, handrails, toeboard, and grating black; nonskid paint usually is gray.

Safety Precautions

Obtain a work permit before doing any work at the plant site. Observe fire precautions when painting in chemical and oil processing plants.

Refer to the discussion in Chap. 3. Read the safety precautions contained in the applicable Paint System Specifications.

Reminders: Close solvent containers. Put all oily rags, paper, and other refuse in closed trash containers. Do not leave paint or equipment unattended on a unit, structure, or a scaffold overnight, as it may be knocked or blown off and create an accident or fire hazard. Clean up daily. Inspect scaffolds, swinging platforms, ropes, and ladders for strength and stability. Use life belts when working on overhead frame work. Wear goggles when doing surface preparation work. Use the safety apparel that is provided.

Notes

The temperature limitations for each of the paint systems given in the applicable PSS must be observed.

Glossary of Terms Used in Maintenance Painting

This glossary is a compilation of terms commonly used in the maintenance painting field, with generally accepted definitions of those terms. Although many of the terms used are generally understood, some are interpreted in several different ways; others are not commonly used because of lack of clear definitions. The purpose of this glossary is to eliminate misunderstanding resulting from conflicting interpretations of these terms and to improve communications between users, applicators, and manufacturers of industrial maintenance coatings.

ASTM CUPS: standard laboratory test cups for measuring viscosity.

ABRASION RESISTANCE: resistance to mechanical wear.

ABRASIVE (see NACE Task Group T-6G-1 Report on "Surface Preparation Abrasives for Industrial Maintenance Painting"[1]): the agent used for abrasive blast cleaning; for example, sand, grit, etc.

ABSORPTION: process of soaking up, or assimilation of one substance by another.

ACCELERATOR: catalyst; a material which accelerates the hardening of certain coatings.

ACETONE: dimethyl ketone; solvent.

ACID NUMBER: a numerical index of free acid in an oil or resin.

ACOUSTIC PAINT: paint which absorbs or deadens sound.

Reprinted from MATERIALS PROTECTION, Vol. 4, No. 1, pp. 73–80, January, 1965, with permission from the National Association of Corrosion Engineers, Houston, Texas.

ACRYLIC RESIN: a clear resin polymerized from acrylic acid and methacrylic acid.

ACTIVATOR: catalyst or curing agent.

ADAPTORS: connectors for joining parts of different sizes.

ADDUCT CURING AGENT: a curing agent combined with a portion of the resin.

ADHESION: bonding strength; the attraction of a coating to the substrate.

ADSORPTION: process of attraction to a surface; attachment; the retention of foreign molecules on the surface of a substance.

AGGLOMERATION: random attachment of single units to form groups; formation of masses of pigments; not dispersed.

AGING: remaining undisturbed.

AGITATOR: stirrer; mixer.

AIR ADJUSTING VALVE: spray gun valve controlling input air.

AIR BUBBLE: dry bubble in paint film caused by entrapped air.

AIR CAP (or air nozzle): perforated housing for atomizing air at head of spray gun.

AIR DRYING: dries by oxidation or evaporative drying by simple exposure to air without heat or catalyst.

AIR ENTRAINING AGENTS: natural wood resins, fats, inorganic materials, sulfonated compounds, and oils for air entrapment in concrete up to 10 percent.

AIR ENTRAPMENT: inclusion of air bubbles in paint film.

AIR HOSE: hose of air supply quality, usually red.

AIRJET (sandblasting): a type of sandblasting gun in which the abrasive is conveyed to the gun by partial vacuum.

AIRLESS SPRAYING: spraying without atomizing air, using hydraulic pressure.

AIR MANIFOLD (pig): common air supply for several lines.

AIR TRANSFORMER: device for controlled reduction in air pressure.

AIR VALVE: control valve in air line system.

AIR VOLUME: quantity of air in cubic feet (usually per minute) at normal (atmospheric) pressure.

ALCOHOL: a flammable solvent, miscible with water; alcohols commonly used in painting are ethyl alcohol (ethanol) and methyl alcohol (methanol, wood alcohol).

ALDEHYDES: chemical compounds containing R-CHO grouping.

ALIPHATIC HYDROCARBONS: "straight chain" solvents of low solvent power, derived from petroleum.

ALKALI: caustic; inorganic compounds which release hydroxyl groups in aqueous media.

ALKYD RESINS: resins prepared by reacting alcohols and acids.

ALLIGATORING: surface imperfections of paint having the appearance of alligator hide.

ALLYL RESINS: resins prepared from allyl alcohol.

AMBIENT TEMPERATURE: room temperature or temperature of surroundings.

AMERICAN GALLON: 231 cubic inches.

AMIDES: compounds containing oxygen and amino (NH_2) groupings.

AMINE ADDUCT: amine curing agent combined with a portion of the resin.

AMINES: organic substituted ammonia; organic compounds having an NH_2 group.
AMINO RESINS: those containing reactive NH_2 groups.
AMYL PHENOL RESINS: particular group of organic film formers. See phenolic resins.
ANCHOR PATTERN: profile, surface roughness.
ANGLE BLASTING: blast cleaning at angles less than 90 degrees.
ANGLE OR DEGREE (airless spray cap): orifice angle; controls width of spray pattern angle.
ANHYDRIDE: compound not containing water.
ANHYDROUS: dry; free of water in any form.
ANION: negatively charged ion.
ANNULAR ORIFICE: circular opening.
ANODE: the electrode at which corrosion (oxidation) occurs.
APPLICATOR: one who applies; tool for applying.
ARCING: swinging spray gun away from perpendicular.
ARGILLACEOUS: clay containing.
AROMATIC HYDROCARBONS: ring compounds; strong solvents.
ASPHALT: residue from distilling petroleum; also a natural complex hydrocarbon found in Trinidad, the United States and elsewhere.
ASPHALT CUT BACK: asphalt plus thinner; asphalt solution; asphalt coating formed by dissolving asphalt.
ASPHALT EMULSION: asphalt dispersion; not a solution; a water emulsion of asphalt.
ASPHALT IMPREGNATED: containing absorbed asphalt.
ATOMIZE: break stream into small particles.
AURAND SCALER: proprietary cleaning tool using cutter wheel bundles.
BAKING FINISH: finish requiring heat cure.
BANDING: identifying with strips of tape.
BARRIER: shielding or blocking device.
BASE: substrate.
BINDER: resin; film former; vehicle.
BITUMEN: product of asphalt or coal tar origin.
BITUMINOUS COATING: coal tar or asphalt based coating.
BLAST ANGLE: angle of nozzle with reference to surface; also angle of particle propelled from wheel with reference to surface.
BLAST CLEANING: cleaning with propelled abrasives.
BLEACHING: removing color.
BLEEDER GUN: a spray gun with no air valve; trigger controls fluid flow only.
BLEEDING: surface flotation of color from under coats.
BLISTERING: bubbling in dry or partially dry paint film.
BLOCK COAT (barrier coat or transition primer): tie coat (adhesive) between non-compatible paints.
BLOOMING: whitening; moisture blush; blushing.
BLOW-BACK (spray term): rebound.
BLUSHING: whitening and loss of gloss due to moisture; blooming.
BODY: viscosity; middle or under (coat).

BOILERS (solvent): solvents of particular evaporation rate.

BONDERIZING: a five-step proprietary custom process for phosphatizing.

BONDING: adhesion.

BOOMERANG (Mikrotest Gauge): a single magnet proprietary direct reading, dry film thickness gauge.

BOUNCE-BACK: spray rebound.

BOXING: mixing by pouring from one container to another.

BRIDGING: forming a skin over a depression.

BRIGHT BLAST: white blast.

BRITTLENESS: degree of susceptibility to cracking or breaking by bending.

BRONZE TOOLS: non-sparking tools.

BRONZING: formation of metallic sheen on a paint film.

BRUSHABILITY: ease of brushing.

BRUSH-OFF BLAST: poorest defined quality of sandblast surface.

BUBBLING: a term used to describe the appearance of bubbles on the surface while a coating is being applied.

BULKING VALUE: volume per unit weight, usually expressed as gallons per pound.

CAKING: hard settling of pigment from paint.

CALCAREOUS: lime containing.

CALCIMINE: pigment used in white wash.

CASEIN PAINT: water thinned paint with vehicle derived from milk.

CATALYST: accelerator; curing agent; promoter.

CAT-EYE: hole or holiday shaped like a cat's eye; cratering.

CATHODE: the electrode at which corrosion (oxidation) usually does not occur.

CATHODIC PROTECTION: corrosion prevention by sacrificial anodes or impressed current.

CATION: positively charged ion.

CAVITATION: undercutting; crevice forming; may be caused by fluids at high velocities and by flashing from liquid to gaseous state.

CELLULOSE RESINS: those prepared from cellulose derivatives.

CEMENT FINISHES: coatings containing Portland cement.

CENTIPOISE: a metric unit of viscosity.

CENTRIFUGE: device for separating solids from liquids by centrifugal action.

CHALKING: powdering of surface.

CHAMPAGNE FINISH (effervescence): rapid escape of solvent visible by bubbling.

CHECK: shallow crack of short length.

CHECKING: formation of checks.

CHIPPING: (1) cleaning steel using special hammers. (2) type of paint failure.

CHLORINATED RUBBER: a particular film former used as a binder, made by chlorinating natural rubber.

CLEANER: (1) detergent, alkali, acid or other cleaning material; usually water or steam borne. (2) solvent for cleaning paint equipment.

CLEAN SURFACE: one free of contamination.

COAL TAR: black residue remaining after coal is distilled.

COAL TAR EPOXY PAINT: paint in which binder or vehicle is combination of coal tar with epoxy resin.
COAL TAR URETHANE PAINT: paint in which binder or vehicle is combination of coal tar with polyurethane resin.
COATINGS: surface coverings; paints; barriers.
COAT OF PAINT: one layer of dry paint, resulting from a single wet application.
COBWEBBING: premature drying causing a spider web effect.
COHESION: property of holding self together.
COLD-CHECKING: checking caused by low temperature.
COLD CRACKING: cracking occurring at low temperature.
COLOR DYNAMICS: scientific use of action colors.
COLOR-FAST: non-fading.
COLOR RETENTION: ability to retain original color.
COMMERCIAL BLAST: surface quality between brush-blast and near-white blast finish.
COMPATIBILITY: ability to mix or adhere properly to other components.
COMPOSITION: analysis; make-up.
CONTINUITY: degree of being intact or pore free.
CONTRAST RATIO: the coefficient of reflection of the black surface area divided by the coefficient of reflection of the white area.
CONVERTER: that which causes change to different state; catalyst; curing agent; promoter.
COPOLYMERS: large molecules resulting by simultaneous polymerization of different monomers.
COPPER SULFATE TEST (for mill scale): copper color indicates absence of mill scale when steel is swabbed with 5 to 10 percent solution.
CORROSION: decay; oxidation; deterioration due to interaction with environment; eaten away by degrees.
CORROSION FATIGUE: loss of strength caused by corrosion.
COUMARONE-INDENE RESINS: particular type of organic binder or resin; coal tar resins.
COVERAGE: milage, usually in square feet per gallon for a given dry film thickness.
CRACKING: splitting; disintegration of paint by breaks through film.
CRATERING: formation of holes or deep depressions in paint film.
CRAWLING: shrinking of paint to form uneven surface.
CRAZING: development of non-uniform surface appearance of myriad tiny scales or cracks.
CREEPAGE: see crawling.
CROSS-LINKING: a particular method by which chemicals unite to form films.
CROSS-SPRAY: spraying first in one direction and second at right angles.
CRYSTALLINE STRUCTURE: a structure in which components have a regular pattern of planes.
CURING: setting up; hardening.
CURING AGENT: hardener; promoter.
CURTAINING: sagging.

CURTAINS: sags having a draped effect.
CYCLING (of pump): interval between strokes.
DAMP: wet; not dry.
DEADMAN VALVE: shutoff valve at blast nozzle, operated by remote control.
DECORATIVE PAINTING: architectural painting; aesthetical painting.
DEGREASER: chemical solution (compound) for grease removal.
DELAMINATION: separation of layers.
DENSITY: weight per unit volume.
DETERGENT: cleaning agent.
DETERIORATION: decay.
DEW POINT: temperature at which moisture condenses.
DILUENT: a liquid which lowers viscosity and increases the bulk but is not necessarily a solvent for the solid ingredients; a thinner.
DISCOLORATION: color change.
DISPERSION: suspension of one component in another.
DISTENSIBILITY: ability to be stretched.
DISTILLATION: purification or separation by volatilizing and condensing.
DOCTOR BLADE: knife applicator.
DOLOMITE: carbonate of calcium and magnesium.
DOUBLE REGULATION: regulation of both pot and gun air pressure.
DRIER: chemical which promotes oxidation or drying of paint.
DRIFT: (overspray): spray loss.
DROP (scaffold): one vertical descent of the scaffold.
DROP CLOTH: protective cover.
DRYING OIL: an oil which hardens in air.
DRYING TIME: time interval between application and final cure.
DRY SPRAY: overspray or bounce back; sand finish due to spray particle being partially dried before reaching the surface.
DRY TO HANDLE: time interval between application and ability to pick up without damage.
DRY TO RECOAT: time interval between application and ability to receive next coat satisfactorily.
DRY TO TOUCH: time interval between application and tack-free time.
DULLING: loss of gloss or sheen.
EDGING: striping.
EFFLORESCENCE: deposit of soluble white salts on surface of brick and other masonry.
EGGSHELL: semi-gloss; dull.
ELASTICITY: degree of recovery from stretching.
ELCOMETER: a two-prong proprietary magnetic direct reading dry film thickness gauge.
ELECTROLYSIS: decomposition by means of an electrical current.
ELECTROLYTE: a substance which disassociates into ions when in solution or a fused state and which will then conduct an electric current. Sodium chloride and sulfuric acid are common examples.
ELECTROSTATIC SPRAY: spraying in which electric charge attracts paint to surface.

ELONGATION: stretch.
EMULSION PAINT: water base paint with an emulsified resin vehicle.
ENAMEL: pigmented varnish; any hard, glossy coating.
ENDOTHERMIC: a change or process which takes place with absorption of heat.
EPOXY ADDUCT: epoxy resin having all of the required amine incorporated by requiring additional epoxy resin for curing.
EPOXY AMINE: amine cured epoxy resin.
EPOXY ESTER: epoxy modified oil; single package epoxy.
EPOXY RESINS: film formers usually made from bisphenol and epichlorohydrin.
EROSION: wearing away of paint films; heavy chalking tends to accelerate erosion.
ESTER: reaction product of alcohol and acid; an organic salt.
ESTIMATE: compute; calculated cost of a job.
ETCH: surface attack by chemical means.
EVAPORATION RATE: rate of solvent release.
EVAPORATION RATE, FINAL: time interval for complete evaporation of all solvents.
EVAPORATION RATE, INITIAL: time interval during which low boiling solvent evaporates completely.
EXOTHERMIC: a change or process in which heat is given off.
EXPLOSION: cratering from release of solvent after surface is dry; also see blistering.
EXPLOSIVE LIMITS: a range of the ratio of solvent vapor to air in which the mixture will explode if ignited. Below the lower or above the higher explosive limit the mixture is too lean or too rich to explode. The critical ratio runs from about one to seven percent of solvent vapor by volume at atmospheric pressure.
EXTENDER: filler; cheapener.
EXTENSION GUN: pole gun.
EXTERNAL MIX: spray equipment in which fluid and air join outside of aircap.
FDA: see Food & Drug Administration.
FADEOMETER: device for measuring color retention or fade resistance.
FADING: reduction in brightness of color.
FALLOUT (spray): overspray.
FALSE BODY: thixotropic.
FANNING (spray gun technique): arcing.
FAN PATTERN: geometry of spray pattern.
FAST DRYING: dry for recoat in less than 24 hours, quick hardening paint.
FAT PAINT: too much oil.
FATTY ACID: a component of certain drying oils; vegetable oil derivative.
FEATHER EDGE: tapered edge.
FEATHERING: (1) flickering a gun at the end of each stroke; (2) tapering edge.
FEDERAL SPECIFICATIONS: government specifications for formulations, raw material components, or performance.
FERROUS: iron containing.

FIELD PAINTING: painting at the job site.
FILLER: extender; bulking agent; inert pigment.
FILM BUILD: dry thickness characteristics per coat.
FILM FORMER: a substance which forms a skin or membrane when dried from a liquid state.
FILM INTEGRITY: degree of continuity of film.
FILM THICKNESS GAUGE: device for measuring film thickness above substrate; dry or wet film thickness gauges are available.
FILTER: strainer; purifier.
FINENESS OF GRIND: measure of particle size or roughness of liquid paint; degree of dispersion of pigment in the binder.
FINGERS (airless): broken spray pattern; fingerlike.
FIRE RETARDANT PAINT: a paint which will delay flaming or overheating of substrate.
FISH EYE: see cratering.
FLAKING: disintegration in small pieces or flakes.
FLAME CLEANING: method of surface preparation of steel using flame.
FLAMMABILITY: measure of ease of catching fire; ability to burn.
FLASH POINT: the lowest temperature at which a given flammable material will flash if a flame or spark is present.
FLAT FINISH: dull finish, no gloss.
FLATTING AGENT: paint ingredient causing low gloss.
FLEXIBILITY: ability to be bent without damage.
FLOATING: separation of pigment colors on surface.
FLOCCULATION: see agglomeration.
FLOCKING: a coating process producing velvet-like surfaces.
FLOODING: see floating.
FLOW: a measure of self spreading ability; spread.
FLUID ADJUSTING SCREW: a screw on a spray gun which controls the amount of fluid entering the gun.
FLUID FLOW: a measure of flow through a gun with atomizing air shut off.
FLUID HOSE: specially designed hose for paint materials; usually black.
FLUID NOZZLE: fluid tip or orifice; in a broader sense it connotes needle and tip combination.
FLUID TIP: orifice in gun into which needle is seated.
FOAMING: frothing.
FOGGING: misting.
FOOD & DRUG ADMINISTRATION (FDA): agency involved with linings for food or pharmaceutical service.
FORCED DRYING: acceleration of drying by increasing the temperature above ambient temperature accompanied by forced air circulation.
FORD CUP: a proprietary viscosity measuring device.
FROTHING: foaming.
FUNGICIDE: a substance poisonous to fungi, retards or prevents fungi growth.
FUNGUS: any of a group of plants, such as molds, mildew, mushrooms, smuts, etc.

FURANE RESINS: dark chemical resistant resins made from furfuryl alcohol, furfural and phenol.

FURFURAL: a particular type of aldehyde used to make furane resins.

GALVANIC CORROSION: corrosion of dissimilar metals in electrical contact.

GALVANIZED STEEL: zinc plated steel applied in a molten bath of zinc.

GAS CHECKING: fine checking; wrinkling, frosting under certain drying conditions; said to be caused by rapid oxygen absorption or by impurities in the air.

GEL: a jelly-like substance.

GELLING (gelation): conversion of a liquid to a gel state.

GENERIC: belonging to a particular family.

GILSONITE: a special bitumen; an asphalt found in Utah; one of the purest of natural bitumens.

GLAZING (paint term): application of transparent or translucent pigment on a painted surface to produce certain blended effects.

GLAZING (puttying): setting glass.

GLOSS: sheen; ability to reflect; brightness; lustre.

GLOSS METER: device for measuring sheen or lustre.

GLOSS RETENTION: ability to retain original sheen.

GRAIN: surface appearance, usually of wood.

GRAY BLAST: commercial blast.

GRIND GAUGE (Hegeman): proprietary instrument for measuring smoothness of liquid paint.

GRIT: an abrasive obtained from slag and various other materials.

GROOVING (roofing term): formation of shallow channels.

GROUND WIRE (airless): a wire attached to dissipate electrostatic charge.

GUIDE COAT: a coat similar to the finish or color coat but of a different color to assure good coverage.

GUN DISTANCE: space between tip of gun and work.

HALIDE: a compound containing fluorine, bromine, chlorine or iodine.

HALOGEN: bromine, chlorine, fluorine or iodine.

HARDENER: curing agent; promoter; catalyst.

HARDNESS: the degree a material will withstand pressure without deformation or scratching.

HAZING: clouding.

HEAVY CENTERED PATTERN: spray pattern having most paint in center, less at edges.

HESIOMETER: proprietary device for measuring cohesion and adhesion.

HIDING POWER: ability to obscure substrate.

HIGH BOILING SOLVENT: a solvent with an initial boiling point above 302F (150C).

HIGH BUILD: producing thick dry films per coat.

HIGH FLASH NAPHTHA: an aliphatic solvent having a high flash point (113F, 45C).

HOLD OUT: ability (or property) to prevent soaking into substrate.

HOLIDAY: pinhole; skip, discontinuity; voids.

HOLIDAY DETECTOR: device for detection of pinholes or holidays.

HONEYCOMBING: lack of vertical film integrity; formation of cell structure; voids.

HOSE CLEANER: mechanical device promoting a beneficial swirling action to cleaning solvent.

HOSE RESTRICTION: impediment; reduced diameter.

HOT SPRAY: spraying material heated to reduce viscosity.

HOT SURFACE: above 120 degrees Fahrenheit (48.9C).

HUMIDITY: measure of moisture content; relative humidity is the ratio of the quantity of water vapor in the air to the greatest amount possible at the given temperature. Saturated air is said to have a humidity of 100.

HUMMOCKING (roofing term): formation of raised islands.

HYDRAULIC SPRAYING (see airless): spraying by hydraulic pressure.

HYDROPHILIC: having an affinity for water; capable of uniting with or dissolving in water.

HYDROPHOBIC: having an antagonism for water; not capable of uniting or mixing with water.

HYDROXYL: chemical radical; OH; basic nature.

HYGROSCOPIC: having a tendency to absorb water.

IMPACT RESISTANCE: a measure of resistance to a blow; ability to resist deformation from impact.

INCOMPATIBILITY: inability to mix with or adhere to another material.

INDICATOR (pH) PAPER: a vegetable dyed paper indicating relative acidity or basicity.

INERT: not reactive.

INERT PIGMENT: a non-reactive pigment; filler.

INFLAMMABILITY: measure of ease of catching fire; ability to burn; use of the word flammability is preferred to inflammability due to the possible misinterpretation of the prefix "in" as a negative.

INHIBITIVE PIGMENT: one which retards corrosion process.

INHIBITOR: an agent added to retard corrosion.

INORGANIC: containing no organic carbon.

INORGANIC COATINGS: those employing inorganic binders or vehicles.

INSULATION: thermal, electrical, or sound barrier material; a poor conductor.

INTERCOAT CONTAMINATION: presence of foreign matter between successive coats.

INTERMEDIATE COAT: middle coat; guide coat.

INTERNAL MIX: a spray gun in which the fluid and air are combined before leaving the gun.

INTUMESCE: to form a voluminous char on ignition; foaming or swelling when exposed to flame.

ION: an electrically charged atom or group of atoms.

IRON PHOSPHATE COATING: conversion coating; chemical deposit.

ISOCYANATE RESINS: resins characterized by CNO grouping; polyurethane resins.

JAPAN: dark colored glossy varnish.

JAPAN DRIER: weak mixture of driers.

JEEP TEST: continuity test using low voltage circuit.

KB VALUE: measure of solvent power.

KTA PANEL: a proprietary paint test panel with unique configuration and markings.

KTA RATING SYSTEM: 10 for no failure, 0 for complete failure; proprietary method of measuring paint disintegration over KTA Panels.

KAURI REDUCTION: test for solvent power of petroleum solvents.

KETONES: organic solvents containing CO grouping; commonly used ketones are acetone-dimethyl ketone; MEK-methyl ethyl ketone; and MIBK-methyl isobutyl ketone.

KREB UNITS: units of viscosity.

LACQUERS: coatings which dry by evaporation of solvent.

LAITANCE: milky white deposit on new concrete; efflorescence.

LAMINAR SCALE: rust formation in heavy layers.

LAP: see overlap.

LATEX: rubber like; a common binder for emulsion (water) paints; there are natural and synthetic latexes.

LEACHING: the process of extraction of a soluble component from a mixture with an insoluble component by percolation of the mixture with a solvent, usually water.

LEAFING: orientation of pigment flakes in horizontal planes.

LEVELLING: flowing out to films of uniform thickness; loss of brush marks in paint.

LIFTING: softening and raising of an undercoat by application of a top coat.

LININGS: internal barriers; linings may be coated or sheet type.

LIVERING: formation of curds or gelling.

LONG OIL: a resin having a large quantity of oil cooked per 100 pounds of resin (25 gallons or more per 100 pounds of resin).

LOOSE FLAKE (mill scale): thin, easily-removed scale.

LOW BOILING SOLVENT: a solvent with an initial boiling point below 302F (150C).

LOW PRESSURE SPRAYING: conventional air spraying.

MAC (maximum allowable concentration): maximum allowable concentration in parts of solvent vapor to one million parts of air in which a worker can work not more than eight consecutive hours without an air fed mask; the lower the MAC number, the more toxic the solvent.

MEK: see methyl ethyl ketone.

MIBK: see methyl isobutyl ketone.

MVT: see moisture vapor transmission.

MAINTENANCE PAINTING: (1) repair painting; any painting after the initial paint job; in a broader sense it includes painting of items installed on maintenance; (2) all painting except that done solely for aesthetics.

MALEIC RESINS: a class of resins obtained from polymerization of maleic acid or maleic anhydride with alcohols; rosins; etc.

MANDREL TEST: a physical bending test for adhesion and flexibility.

MASKING: covering areas not to be painted.

MASS TONE: base covering.

MASTIC: a heavy bodied coating of high build.

MELAMINE RESINS: synthetic resins which are condensate products of formaldehyde and melamine; they require baking.

METALLIZING: mechanical deposition of one metal on another.

METHYL ETHYL KETONE (MEK): a strong solvent.

METHYL ISOBUTYL KETONE (MIBK): a strong solvent.

MIKROTEST GAUGE: a proprietary single magnet dry film thickness gauge.

MIL: one one-thousandth of an inch; .001″.

MILAGE: coverage rate; square feet per gallon at a given thickness.

MILDEW: fungus.

MILDEWCIDE: substance poisonous to mildew; prevents or retards growth of mildew.

MILD STEEL: structural steel; SAE 1020.

MILL SCALE: oxide layer formed on steel by hot rolling.

MILL SCALE BINDER: gray oxide layer between mill scale and steel.

MILL WHITE: one coat high hiding power interior paint.

MINERAL SPIRITS: aliphatic solvent with solvency similar to naphtha.

MISCIBLE: capable of mixing or blending uniformly.

MISSES: holidays; skips; voids.

MIST-COAT: thin tack coat; thin adhesive coat.

MOISTURE AND OIL SEPARATOR: trap on air compressor or in air lines.

MOISTURE VAPOR TRANSMISSION (MVT): moisture vapor transmission rate through a membrane; also see perm.

MONOMER: composed of single molecules; a basic chemical used to make polymers.

MOPPING: swabbing, as with roofing asphalt.

MOTTLING: speckling; an uneven color on paint.

MUD-CRACKING: irregular cracking, as in a dried mud puddle.

MULTICOLOR FINISHES: speckled finishes; paints containing flecks of colors different from the base color.

NAPHTHA: an aliphatic solvent cut; hydrocarbons of the CnHn+2 series.

NEAR-WHITE BLAST CLEANING: surface preparation nearly equal to white-blast finish.

NEEDLE (spray gun): fluid metering pin.

NEOPRENE: a rubber-like film former; a type of elastomer based on polymers of 2-chloro-butadiene-1,3.

NONDRYING OIL: one which will not harden in air.

NONFERROUS: containing no iron.

NONFLAMMABLE: incombustible.

NONTOXIC: not poisonous.

NONVOLATILE: solid; nonevaporating; the portion of a paint left after the solvent evaporates.

NOZZLE: orifice; sandblast nozzle; spray gun nozzle.

NYLON RESINS: a particular group of film formers having recurring amide groups—CONH. as an integral part of the main polymer chain; polyamide resins.

OIL ABSORPTION: a measure of the ability of pigments to absorb oil.
OIL COLOR: coloring (pigment or dye) dispersed in oil.
OIL LENGTH: gallons of oil reacted with 100 pounds of resin.
OLEORESINOUS: film former containing oil and resin.
OPACITY: hiding power.
ORANGE PEEL: dimpled appearance of dried film; resembling orange peel.
ORGANIC: containing carbon.
ORGANOSOL: film former containing resin plasticizer and solvent; colloidal dispersion of a resin in plasticizer containing more than 5 percent volatile content.
ORIFICE: opening; hole.
OSMOSIS: transfer of liquid through a paint film or other membrane.
OSMOTIC BLISTERING: formation of blisters containing liquid.
OVERATOMIZED: dispersed too finely by use of excessive atomizing air pressure.
OVERCOAT: second coat; top coat.
OVERLAP: portion (width) of fresh paint covered by next layer.
OVERSPRAY: sprayed paint which did not hit target; waste.
OXIDATION: combination with oxygen; drying; burning; rusting.
OXIDE: chemical compound of an element, usually a metal, with oxygen.
PVA: see polyvinyl acetate.
PVC: see pigment volume concentration or polyvinyl chloride.
PAINT HEATER: device for lowering viscosity of paint by heating.
PAINT PROGRAM: comprehensive painting plan. (see NACE Task Group T-6D-3 Report "Industrial Maintenance Painting Program"[3]).
PAINT PROJECT: single paint job.
PAINT SYSTEM: the complete number and type of coats comprising a paint job. In a broader sense, surface preparation, pretreatments, dry film thickness, and manner of application are included in the definition of a paint system.
PARTICLE SIZE DISTRIBUTION: percentages of particles of different screen sizes.
PASS (spray): motion of the spray gun in one direction only.
PASSIVATION: act of making inert or unreactive.
PATTERN LENGTH: height of spray pattern.
PATTERN WIDTH: width of spray pattern at vertical center.
PEELING: failure in which paint curls from substrate.
PERM: a unit for expressing MVT rate; a perm-inch = 1 grain of water per hour per square foot per one inch thickness (except where otherwise noted) per one-inch difference in mercury vapor pressure on each side of membrane.
PERMEABILITY: quality or state of being permeable.
PHENOLIC RESINS: particular group of film formers, phenol-formaldehyde type.
PHOSPHATIZE: form a thin inert phosphate coating on surface usually by treatment with $H_3 PO_4$ (phosphoric acid).
PHTHALIC RESINS: a particular group of film formers; alkyds.
pH VALUE: measure of acidity or alkalinity; pH 7 is neutral; the pH value of acids ranges from 1 to 7, and of alkalis (bases) from 7 to and including 14.

PICKLING: a dipping process for cleaning steel and other metals; the pickling agent is usually an acid.

PIG: see air manifold.

PIGMENT GRIND: dispersion of pigment in a liquid vehicle.

PIGMENTS: solid coloring agents.

PIGMENT VOLUME CONCENTRATION (PVC): percent by volume occupied by pigment in dried film.

PIG TAIL: finger-like spray pattern.

PIN-HOLING: formation of small holes through the entire thickness of coating; see cratering.

PITTING: formation of small, usually shallow depressions or cavities.

PLASTICIZER: a paint ingredient which imparts flexibility.

PLASTISOL: film former containing resin and plasticizer with no solvents.

POCK MARKS: pits; craters.

POLE-GUN: spray gun equipped with an extension tube.

POLYMER: the product of polymerization; large molecules.

POLYMERIZATION: formation of large molecules from small ones.

POLYVINYL ACETATE (PVA): a synthetic resin used extensively in emulsion (water) paints; produced by the polymerization of vinyl acetate.

POLYVINYL CHLORIDE (PVC): a synthetic resin used in solvent type coatings and fluid bed coatings, produced by the polymerization of vinyl chloride; PVC is also used in emulsion (water) paints.

POLYVINYL CHLORIDE ACETATE: a combination of PVA and PVC used in coatings.

POROSITY: hole; degree of integrity or continuity.

POT-LIFE: time interval after mixing during which liquid material is usable with no difficulty.

PRECIPITATION: settling out of solid material.

PRESSURE BALANCE: in spray painting, relationship of pot pressure to atomizing air pressure.

PRESSURE DROP: loss in pressure due usually to length or size of line or hose.

PRESSURE FEED: fluid flow caused by application of air or hydraulic pressure on paint.

PRESSURE FEED PAINT TANK (pressure pot): fluid container in which fluid flow is caused by air pressure.

PREVENTIVE MAINTENANCE PAINTING: spot repair painting; touch up or full coats of paint before rusting starts.

PRIME COAT: first coat.

PRIMER: material used for prime coat.

PRODUCTION RATE (sq ft/day): measurement of surface area cleaned or coated in one working day by one man.

PROFILE: surface contour as viewed from edge.

PROFILE DEPTH: average distance between tops of peaks and bottom of valleys on the surface.

PROPRIETARY: available on the open market under a brand name.

PROTECTIVE LIFE: interval of time during which a paint system protects substrate from deterioration.

PULSATION: surging.
PUMP BYPASS: recirculating line which returns fluid to container.
PUMP RATIO: multiplier of input pressure which indicates output pressure; ratio of air piston area to fluid piston area.
QUICK RELEASE FITTINGS: snap-lock fittings.
REACHING (spray gun): extending spray stroke too far.
REBOUND: paint spray deposit bounced back.
RECOAT TIME: time interval required between application of successive coats.
RED LABEL GOODS: flammable or explosive materials with flash points below 80 F. (26.7 C).
REDUCER: a material which lowers viscosity but is not necessarily a solvent for the particular film former; thinner.
REFLECTANCE: degree of light reflection.
REPAINTING: repetition of a complete painting operation including surface preparation.
RESIN: a material, natural or synthetic, contained in varnishes, lacquers, and paints; the film former.
RESPIRATOR: safety breathing mask.
RETICULATION: a surface defect of net-like appearance.
RISE: height.
ROLLER COATING: the act of, or the material, applied with a roller.
ROUND PATTERN: circular spray pattern.
RUNS: curtains; sags.
RUST: corroded iron; red iron-oxide; also other metal oxides formed by corrosion.
RUST BLOOM: discoloration indicating the beginning of rusting.
SSPC: Steel Structures Painting Council.
SAFETY VALVE: pressure release valve preset to safe operating limit.
SAGS: runs.
SALT SPRAY: a salt fog test environment.
SANDBLAST: blast cleaning using sand as an abrasive; for different grades of blast cleaning finishes see brush blast, commercial blast, near-white blast, and white blast.
SANDY FINISH: a surface condition having the appearance of sandpaper; overspray.
SATURANT: that substance, usually a liquid, which saturates something else.
SATURATED: holding the maximum amount of saturant it is capable of holding.
SCALE: laminar rust.
SCALER: a hand cleaning chisel.
SCALING: process of delamination.
SEALER: a low viscosity (thin) liquid applied before priming wood or masonry.
SEEDING: formation of small agglomerates.
SEPARATION: division into components or layers by natural causes.
SETTLING: caking; sediment.
SHADE: degree of gray tone in a color.
SHELF LIFE: maximum interval in which a material may be stored in usable condition.

SHELLAC: a resin secreted by insects; a lacquer; resin in alcohol.

SHIELDING: protecting; protective cover against mechanical damage.

SHOP COAT (prime) (see NACE Task Group T-6D-3 report on "Industrial Maintenance Painting Program"[3]): first coat applied in fabricating shop.

SHORT OIL: a varnish prepared by cooking a relatively small quantity of oil with 100 pounds of resin; quick drying; brittle; less than 25 gallons of oil per 100 pounds of resin.

SHOT BLASTING: blast cleaning using steel shot as the abrasive.

SHRINKAGE: decrease in volume on drying.

SILICATE PAINTS: those employing silicates as binders.

SILICONE RESINS: a particular group of film formers; used in water proof and high temperature paints; organosiloxane polymers; semi-organic polymers containing silicon.

SILKING: a surface defect characterized by parallel hairlike striations in coated films.

SKINNING: formation of a solid membrane on top of a liquid.

SKIPS: holidays; misses; uncoated area; voids.

SLOW DRYING: requiring 24 hours or longer before recoating.

SLUG: surge of material; blob.

SOLID: non-volatile portion of paint.

SOLIDS VOLUME: percentage of total volume occupied by non-volatiles.

SOLUTION: a liquid in which a substance is dissolved.

SOLVENCY: measure of ability to act as a solvent.

SOLVENT: a liquid in which another substance may be dissolved.

SOLVENT BALANCE: ratio of amounts of different solvents in a mixture of solvents.

SOLVENT POP: blistering caused by entrapped solvent.

SOLVENT POWER: see solvency.

SOLVENT RELEASE: ability to permit solvents to evaporate.

SOLVENT WASH: cleaning with solvent.

SPALLING: the cracking, breaking or splintering of materials, usually due to heat.

SPARK-PROOF TOOLS: bronze beryllium tools.

SPARK TESTING: detection of holidays (flaws) using electric spark.

SPECIFIC GRAVITY: ratio of weight of a given volume to weight of an equal volume of water at the same temperature.

SPECULAR GLOSS: mirror-like reflectance.

SPEWING: irregular or intermittent surging with subsequent liquid spillage.

SPIDER (power staging): a proprietary mechanical boatswain's chair or platform.

SPIT: sputter.

SPOT REPAIR: preventive maintenance; repainting of small areas.

SPRAY CAP: front enclosure of spray gun equipped with atomizing air holes.

SPRAY HEAD: combination of needle, tip, and air nozzle.

SPRAY PATTERN: configuration of spray, gun held steady.

SPREADING RATE: coverage, milage, usually at specified dry thickness.

SPUTTERING FLOW: spitting, surging.

SQUEEGEE: rigid bar applicator.
STATIC WIRE: ground wire.
STEAM CLEAN: a cleaning process using live steam.
STRAIN: to filter.
STREAKS: a surface defect characterized by essentially parallel lines of different colors or shades.
STRIPPING: edge painting prior to priming.
STROKE (spray): a single pass in one direction.
STYRENE-BUTADIENE: resin; copolymer of styrene and butadiene.
SUBSTRATE: surface to be painted.
SUCTION FEED (sandblaster): one in which the abrasive is syphoned to the nozzle.
SUCTION FEED (spray gun): one in which the fluid is syphoned to the spray head.
SURFACER: a paint used to smooth the surface before finish coats are applied.
SURFACE TENSION: cohesive force on liquid surface.
SURGE: see spewing; non-continuous flow.
SURGE CHAMBER (airless spray): a device to eliminate uneven fluid flow.
SWEATING: condensing moisture on a surface.
SWELLING: increasing in volume.
SWIVEL FITTING: one capable of being moved in any direction.
SWIVEL HEAD: spray head adjustable to deliver spray in many directions.
SYNTHETIC: manufactured; not occurring naturally.
TLV (Threshold Limit Value): see MAC.
TACK: degree of stickiness.
TAIL (solvent evaporation): highest boiling solvent fraction.
TAIL LINE: short piece of blast hose smaller than the main hose to permit better maneuverability.
TAILS (airless spray): finger-like spray pattern.
TANK WHITES: good hiding; self cleaning; white paints; usually alkyds.
TAPERED PATTERN: elliptical shaped spray pattern; a spray pattern with converging lines.
TAPE TEST: a particular type of adhesion test.
TENACITY: ability to stick together; cohesiveness; adhesiveness.
TENSILE STRENGTH: resistance to elongation; the greatest longitudinal stress a substance can bear without rupture or remaining permanently elongated.
TERPENE RESINS: a particular group of film formers, prepared from isomeric hydrocarbons such as turpentine or similar oleo-resins.
TEST PATTERN: spray pattern used in adjusting spray gun.
THERMOPLASTIC: mobile or softens under heat.
THERMOSETTING: becomes rigid under heat and cannot be remelted.
THINNERS: volatile organic liquids for reducing viscosity; solvents.
THIXOTROPIC: false-bodied; a gel which liquifies with agitation but gels again on standing.
TINSLEY GAUGE: a proprietary pencil-like, single magnet, dry film thickness gauge.

TINT: degree of white in a color; a color produced by the mixture of white paint or pigment with a non-white colored paint or pigment.

TONER: a color modifier.

TOOTH: profile; mechanical anchorage; surface roughness.

TOP COATING: finish coat.

TOUCH-UP PAINTING: spot repair painting usually conducted a few months after initial painting.

TOXIC: poisonous.

TOXICITY: degree of poisonousness or harmfulness.

TRANSITION PRIMER (block or barrier coat): coating compatible with primer and also with a finish coat which is not compatible with primer.

TRIGGER: operating lever of spray gun.

TRIGGERING: intermittent squeezing and releasing of trigger.

TUBERCULE: nodule; pimple.

TWO-COMPONENT GUN: one having two separate fluid sources leading to spray head.

UNDERATOMIZED: not dispersed or broken-up fine enough.

UNIT COST: cost per given area.

UREA FORMALDEHYDE: a particular group of film formers; usually requires baking; produced by reacting urea with formaldehyde.

UREA MELAMINE: see melamine.

UREA RESINS: a particular group of film formers (amino resins).

URETHANE RESINS: a particular group of film formers; isocyanate resins.

VM&P NAPHTHA: varnish and paint manufacturers naphtha; an aliphatic solvent.

VAPOR DEGREASING: a cleaning process utilizing condensing solvent as the cleaning agent.

VAPORIZATION: conversion from liquid or solid to a gaseous state; phase change.

VARNISH: liquid composition of oil, resin, thinners and driers, which is converted to a transparent or translucent solid film after application as a thin layer or coat.

VEHICLE: liquid carrier; binder; anything dissolved in the liquid portion of a paint is a part of the vehicle.

VEILING: curtaining; sagging.

VENTURI: a tube having a restriction to promote velocity increase.

VERTICAL PATTERN: a spray pattern whose longest dimension is vertical.

VINYL ACETATE: a particular resin monomer; obtained by reaction of acetylene and acetic acid; see PVA, polyvinyl acetate.

VINYL CHLORIDE: a particular resin monomer; obtained by reaction of acetylene and hydrochloric acid, cracking of ethylene dichloride, or reaction of ethylene dichloride and caustic soda; see PVC, polyvinyl chloride.

VINYL COATING: one in which the major portion of binder is of the vinyl resin family.

VINYL COPOLYMER: resins produced by copolymerizing vinyl acetate and vinyl chloride.

VINYL RESINS: a particular group of film formers; see PVA and PVC.

VISCOSITY: a measure of resistance to flow.
VISCOSITY CUP: a device for measuring viscosity.
VOIDS: holidays, holes, skips.
VOLATILE CONTENT: percentage of materials which evaporate.
VOLATILES: fluids which evaporate rapidly.
WASHING: erosion of a paint film after rapid chalking.
WASH PRIMER: a thin inhibiting paint usually chromate pigmented with a polyvinyl butyrate binder.
WATER BLASTING: blast cleaning using high velocity water.
WATER SPOTTING: a surface defect caused by water droplets.
WEATHEROMETER: a testing device intended to simulate atmospheric weathering.
WELD JOINTS: beads of weld joining two members.
WELD SLAG: amorphous deposits formed during welding.
WELD SPATTER: beads of metal left adjoining a weld.
WELD SPLATTER: see weld spatter.
WET EDGE: fluid boundary.
WET FILM GAUGE: device for measuring wet film thickness.
WET FILM THICKNESS: thickness of liquid film immediately after application.
WET SPRAY: spraying so that surface is covered with paint that has not started to dry.
WETTING STRENGTH: the maximum distance or penetration the vehicle is capable of delivering the paint or coating assembly in a vertical or horizontal direction on a specific substrate.
WETTING TIME: the time required for a vehicle to reach the end point of distance and penetration on a metal.
WHIP BLAST: see brush blast.
WHIP LINE: see tail line.
WHIPPING (spray gun): arcing, waving.
WHITE BLAST: highest degree of blast cleaning.
WHITING: Paris white; gliders white; fine ground, naturally occurring calcium carbonate, $CaCO_3$, about 98 percent pure. Used as an inexpensive filler and extender.
WICKING: absorption of liquid by capillary action.
WIRE BRUSH: a hand cleaning tool comprised of bundles of wires; also the act of cleaning a surface with a wire brush, including power brushes.
WRINKLING: a surface defect resembling the skin of a prune.
WRIST ACTION (spray gun): swiveling of wrist without arcing forearm.
YELLOWING: development of yellow color or cast, in whites, on aging.
ZINC PHOSPHATE COATING: a thin, inorganic deposit formed on zinc treated with phosphoric acid.
ZINC SILICATE: inorganic zinc coating.
ZINC YELLOW: zinc chromate.

References

1. Surface Preparation Abrasives for Industrial Maintenance Painting. A Report of NACE Technical Unit Committee T-6G, Surface Preparation for Protective

Coatings. Prepared by Task Group T-6G-1 on Abrasive Blast Cleaning Media for Surface Preparation in conjunction with Task Group T-6D-13 on Surface Preparation Media, *Materials Protection,* **3,** 76 (1964) July.

2. Proposed Definitions of Sandblast Cleaned Surfaces. A Report of NACE Technical Committee T-6G, Surface Preparation for Protective Coatings. Prepared by Task Group T-6G-2 on Definitions of Degrees of Blast Cleaning Surface Preparation for Organic Coatings. Available from NACE; to be published soon.
3. Industrial Maintenance Painting Program. Report of NACE Technical Unit Committee T-6D. *Corrosion,* **16,** 301t (1960) June.

Bibliography

Air Pollution Control District: "Rule 66: Organic Solvents; Rule 66.1: Architectural Coatings; Rule 66.2: Disposal and Evaporation of Solvents," City of Los Angeles, Calif., July, 1966.

American Conference of Governmental Industrial Hygienists: "A Guide for Uniform Hygiene Codes or Regulations for Industrial Spray Coating," Supplement #4, ACGIH, Cincinnati, Ohio, Sept., 1958.

American Conference of Governmental Industrial Hygienists: "Threshold Limit Values for 1968," ACGIH, Cincinnati, Ohio, 1968.

Armitage, F.: "The British Paint Industry," Pergamon Press, Long Island City, N.Y., 1967.

Associated Equipment Distributors: "Compilation of Nationally Averaged Rental Rates and Model Reference Data for Construction Equipment," 20th ed., AED, Oak Brook, Ill., 1969.

Bacon, G., and R. Volkening: "Color Can Revolutionize Your Plant," *Petroleum Refiner,* Jan. 1957.

Bay Area Air Pollution Control District: "Regulation 3—The Basic Idea," *Air Currents,* Vol. 10, No. 1, City of San Francisco, Calif., Jan. 1967.

Bay Area Air Pollution Control District: "Regulation 3," City of San Francisco, Calif., Jan. 1967.

Burns, R. M., and W. W. Bradley: "Protective Coatings for Metals," 3rd ed., Reinhold Publishing Corporation, New York, 1967.

Castleberry, J. R.: "Corrosion Prevention for Structural Steel in the Construction Industry," *Materials Protection,* Vol. 7, No. 1, pp. 19–23, Jan., 1968.

Chicago Pneumatic Tool Company: "Bulletin SP 3490: Industrial Air Tool Safety," Chicago Pneumatic Tool Company, New York, 1967.

Clark, George: "New Corrosion Engineer—What He Should Know About Coatings," *Materials Protection,* Vol. 5, No. 4, pp. 17–19, April, 1966.

Compressed Air and Gas Institute: "Compressed Air and Gas Handbook," CAGI, New York, 1966.

Danziger, Glenn N., and Frederick C. Kinsler: "Formulation of Organic Coatings," D. Van Nostrand Company, Inc., Princeton, N.J., 1967.

Editorial: "The Fight for Clean Air: Respiratory Protective Equipment," *Occupational Hazards,* Vol. 29, pp. 60–63, April, 1967.

Ellis, W. H.: "Mineral Spirits and Rule 66," *Western Paint Review,* Vol. 53, No. 7, pp. 13A–20A, July, 1967.

Evans, U.: "An Introduction to Metals Corrosion," E. Arnold Company, London, 1948.

Fontana, Mars G., and N. D. Greene: "Corrosion Engineering," McGraw-Hill Book Company, New York, 1967.

Gaynes, Norman I.: "Formulation of Organic Coatings," D. Van Nostrand Company, Inc., Princeton, N.J., 1967.

Helms, F. Parker: "To the New Corrosion Engineer in Industrial Coatings: Importance of Painting Economics," *Materials Protection,* Vol. 5, No. 4, pp. 3–4, April, 1966.

Industrial Commission: "Rules of the Industrial Commission: Safety, Dusts, Fumes, Vapors, and Gases; Flammable Liquids," Wisconsin (State) Industrial Commission, Madison, Wis.

Keane, John D., ed.: Steel Structures Painting Manual, Vol. 1: "Good Painting Practice"; Vol. 2: "Systems and Specifications," Revised ed., Steel Structures Painting Council, Pittsburgh, Pa., 1966.

Lichtenstein, Stanley: "The Many Faces of Corrosion," STR-3454, U.S. Department of Commerce, National Bureau of Standards, Washington, D.C., Oct., 1966.

Martens, Charles R., ed.: "Technology of Paints, Varnishes, and Lacquers," Reinhold Book Corporation, New York, 1968.

Martinson, C. R., and C. W. Sisler: "Industrial Painting: The Engineered Approach," Reinhold Book Corporation, New York, 1961.

Myers, Raymond R., and J. S. Long: "Film-Forming Compositions" (in three parts); "Characterization of Coatings" (in two parts), Marcel Dekker, Inc., New York, 1967, 1968, 1969.

NACE Publication: "Corrosion Control Makes Dollars and Sense," a compilation of articles previously published in *Materials Protection,* National Association of Corrosion Engineers, Houston, Tex., Oct., 1967.

NACE Publication 6D461: "The Economics of Chemical Plant Maintenance Painting," *Corrosion,* Vol. 17, No. 12, pp. 115–117, Dec., 1961.

NACE Publication 6D168: "Contract and Plant Force Painting: Advantages and Disadvantages," *Materials Protection,* Vol. 7, No. 2, pp. 39–42, Feb., 1968.

NACE Publication S3-1: "Report on Surface Preparation of Steels for Organic and and Other Protective Coatings," *Corrosion,* Vol. 9, No. 10, pp. 173–185, Oct., 1952.

NACE Publication 57-10: "Protective Coatings for Atmospheric Use: Their Surface Preparation and Application Requirements, Physical Characteristics and Resistances," *Corrosion,* Vol. 13, No. 3, pp. 85–93, March, 1957.

NACE Publication TR-166: "Surface Preparation Abrasives for Industrial Maintenance Painting," *Materials Protection,* Vol. 3, No. 7, pp. 76–80, July, 1964.

NACE Technical Committee Report: "A Manual for Painter Safety," National Association of Corrosion Engineers, Houston, Tex., 1963.

National Fire Protection Association: NFPA No. 33: "Spray Finishing," 1966 ed., NFPA International, Boston, Mass., 1966.

National Research Council: "Publication 653: Field Applied Paints and Coatings," National Academy of Sciences, Washington, D.C., 1959.

Nee, J. W.: "Some Developments in Protective Coatings," *Materials Protection,* Vol. 5, No. 4, pp. 27–29, April, 1966.

Nylen, Paul, and Edward Sunderland: "Modern Surface Coatings," Interscience Publishers, a division of John Wiley & Sons, Ltd., London, 1965.

Painting and Decorating Contractors of America: "Estimating Guide," 1st ed., PDCA, Chicago, Ill., 1968.

Parker, Dean H.: "Principles of Surface Coating Technology," Interscience Publishers, a division of John Wiley & Sons, Inc., New York, 1965.

Parks, L. E.: "Protective Coatings—New Trends in Materials and Applications," *Materials Protection,* Vol. 6, No. 12, pp. 37–39, Dec., 1967.

Reynolds, Raymond H., N. W. Karr, and Karl Buss: "Painting and Decorating Craftsman's Manual and Textbook," 4th ed., Painting and Decorating Contractors of America, Inc., Chicago, Ill., 1965.

Sargent, Walter: "The Equipment and Use of Color," Dover Publications, Inc., New York, 1964.

Shreir, L. L., ed.: "Corrosion: Vol. I, Vol. II," John Wiley & Sons, Inc., New York, 1963.

Speller, F. N.: "Corrosion Causes and Prevention," 3rd ed., McGraw-Hill Book Company, New York, 1951.

Titanium Pigment Corporation: "The Handbook," Titanium Pigment Corporation, Subsidiary of National Lead Company, New York.

Uhlig, Herbert H.: "The Corrosion Handbook," 9th printing, John Wiley & Sons, Inc., New York, 1966.

Utah State Division of Health, Section of Industrial Hygiene: "Useful Criteria in the Identification of Certain Occupational Health Hazards," Utah State Division of Health, Salt Lake City, Utah.

Weaver, Paul E.: "Industrial Maintenance Painting," 3rd ed., National Association of Corrosion Engineers, Houston, Tex., 1967.

Index

Index